Construction Cost Control

National Association of Home Builders
15th and M Streets, N.W.
Washington, D.C. 20005

Construction Cost Control
ISBN 0-86718-153-2
Copyright © 1975, 1978, 1982, 1987 by the
National Association of Home Builders
of the United States
15th and M Streets, N.W.
Washington, D.C. 20005

All rights reserved. No part of this book may be reproduced or utilized in any form or by any means, electronic or mechanical, including photocopying and recording, or by any information storage and retrieval system without permission in writing from the publisher.

When ordering this publication, please provide the following information:

    Title
    ISBN 0-86718-153-2
    Price
    Quantity
    NAHB membership number (as it appears on the *Builder* or *Nation's Building News* label)
    Mailing address (including street number and zip code)

Contents

Introduction	5
Guidelines	5
Production Planning and Control	7
Principles	7
Company Goals	7
Scheduling	8
Supervision	21
Organization	22
Subcontracting	25
Quality Control	28
Design For Production	30
Principles	30
Case Studies	33
Material Cost Control and Reduction	35
Principles	35
Purchasing	35
Material Handling	40
Scrap and Waste	40
Material Use	42
Labor Cost Control and Reduction	50
Preplanning	51
Crew Composition and Size	51
Rework	52
Equipment Utilization	57
Methods	57
Actual Onsite Industrial Engineering Studies	59
Other Labor Cost Control and Savings Ideas	63

Preface

This *Construction Cost Control* is designed to provide information to home and apartment builders, remodelers, and light commercial/industrial builders on field-tested methods of controlling and reducing direct construction costs.

This monumental subject is presented in the context of a series of top management alternatives for decision and action. The ideas, suggestions, and examples set forth are based on the assumption that only minor changes to the building organization, production system, or design and construction of the dwelling or building would be needed to put the control and savings ideas into practice. In short, the ideas and methods in the manual assume that the building organization is essentially fixed, that the method of doing business and producing the dwelling or building is not going to be changed to a major degree, and that no design modification and no major change in the general appearance is going to be required to achieve the desired cost controls and savings.

Many builders will find that they have already put into practice a number of the cost control and savings methods discussed in this book. However, many builders will find a few ideas in each major section that will be new and useful to them.

The contents of the manual are drawn primarily from the results of practical, onsite studies and analyses of various home building operations by the Research Foundation over a period of many years. In addition, many individuals and firms have contributed information used in the manual. Accordingly, the authors wish to acknowledge the major assistance and contribution of numerous home builders whose ideas and both successful and unsuccessful experiences are included, the NAHB Builder Services and Technical Services staff and the technical personnel of a number of home building industry trade associations and manufacturers. While acknowledging the advice and assistance of all these individuals and organizations, it should be made clear that this in no way implies their endorsement of the manual, since the statements, conclusions and tables, etc., are those of the principal authors, E. Lee Fisher, Director, Industrial Engineering Services, and Ralph J. Johnson, President.

NAHB National Research Center

Introduction

There are four major components of a complete cost control and reduction program:
1. Production planning and control
2. Design for production
3. Material cost control and reduction
4. Labor cost control and reduction

For the best results, an overall cost control and reduction program will include all four of the components. Efficient use of labor and materials is difficult to maintain without good production planning. The design of a dwelling unit often dictates the degree of labor and material efficiency. Outdated or inefficient labor methods can easily create production planning and control problems.

This manual presents principles of cost control and reduction for each of the four major areas. Many onsite industrial engineering studies, reviews of successful building operations and discussions with builders were used to develop the principles presented in the manual.

Guidelines

Production Planning and Control

Long-range company goals often dictate how the construction schedule should be developed. Regardless of the type of schedule used, it is the tool by which the job is organized to make things happen at the construction site. The key to translating the construction schedule into meaningful activity is good site supervision. Most builders agree that without good site supervision, the most outstanding scheduling system is worthless. Company organization dictates who is responsible for what and provides a system for fixing responsibility and authority. Because most builders subcontract much of the work, subcontractors should be considered a part of the total builder's organization. Selection of subcontractors and supervision of their work is an important part of production planning and control. Quality control is also an important part of cost control because callbacks and repairs are usually more costly than doing things right the first time.

Design for Production

Cost effective construction usually starts at the drawing board. Within marketing limits, each portion of the dwelling unit should be analyzed for simplicity and standardization. Often, minor design changes can result in substantial reduction of labor and materials. The most experienced and efficient tradesmen cannot entirely compensate for inefficient dwelling designs.

Material Cost Control and Reduction

Material cost control starts with purchasing. Many builders have found that a formal purchase order system can result in increased profits. Supplier selection and determination of what building materials are to be used are such important functions that, in many building firms, purchasing is handled by the builder or by someone else in top management. After materials are delivered, the methods of handling materials are important. The objective of good material handling is to reduce time spent handling materials. Also, good material handling practices often result in a reduction of breakage and pilferage. Control of scrap and waste is an important source of cost reduction. Salvage value of materials often more than compensates for salvage labor costs. Proper engineering can result in a reduction of materials put in place. Materials added after a desired performance level is reached add nothing but excessive costs.

Labor Cost Control and Reduction

By preplanning daily work activities of his crew, the working foreman, or crew leader, can be a prime source of labor cost reduction. Proper crew size and composition can reduce waiting time, thereby improving productivity. Some builders have found several small crews are usually more efficient than

Production Planning and Control

Principles

Principles underlying production planning and control combine, in the best possible way, the resources at the builder's disposal. Planning functions to analyze design requirements and to integrate money, men, machines, and materials to produce the builder's product when the customer wants it.

Production planning and control should help the builder coordinate production with sales, establish and maintain financial plans, and maintain relatively uniform employment levels in relation to sales. By-product benefits include reduction of labor and material costs and usually more efficient use of operating capital.

All resources are considered in production planning and control. The checklist includes:

- Material costs—Both availability and cost vs. value analyses are important. The ideal material this month may not be selected next month.
- Crew characteristics—The skills, abilities and relative costs of available crew members are important factors in make-or-buy decisions as well as in construction system selections.
- Subcontractor capabilities and bidding practices—This knowledge is necessary for make-or-buy decisions and to select preferred subcontractors capable of performing well under the production plan.
- Suppliers—Lead times, prefab capabilities, discounts, transportation, all enter into make-or-buy and construction system decisions.
- Equipment and tooling—These influence determination of onsite or offsite fabrication and structural options.
- Predicted Sales Volumes—Volume of production affects decisions about subcontracting, equipment, production systems, crew balance, and many other production decisions.

Designs are analyzed in terms of all available resources during the planning phase. Feedback from sales or production may cause designs to be revised. At all times, of course, customer preference is kept in mind. If clear customer preferences or local norms are not followed, savings generated by the proposed deviation need to be evaluated against the likely effect on sales and sales costs.

Company Goals

All firms have some kind of established business goals. They may consist of employee's guesses at what the boss wants or they may be set forth in expensively printed manuals. They can be formulated by intuitive feelings or by formal committees. In order to benefit from production planning and control, the company must make business goals clear to those assigned the responsibility of reaching them.

Long-range company goals are based on the firm's current position—"where the firm is today." They involve a careful inventory of that position, including consideration of organizational structure, policies, people and their abilities, production, product quality, financial status, and even public image. An honest inventory will reveal strengths and weaknesses of the existing company organization. A probing search may turn up some disquieting weaknesses as well as unsuspected strengths. The best projected long-range plan is designed to utilize the strengths and strengthen the weaknesses of the firm. While this may be a complicated and time-consuming process in a large firm, an hour or two of quiet thought—pencil in hand—may accomplish much for the small firm.

Another consideration in defining the present business status is the market. Who is the customer? Which age groups are buying? What economic groups are buying? What brings the buyer to the firm? What does he really want to buy? Is he buying living space, a way of life, or prestige?

In assessing the present business means considering items such as product (home) quality, customer's idea of value, and how best to fulfill them, whether related to prestige, durability, luxury, or whatever.

Long-range plans should be sufficiently flexible to allow the firm to go with prevailing currents, but in the desired direction.

Questions that may be helpful in projecting company goals include:
- What can be done if inflation continues to increase building costs?
- What markets are not being adequately served today?
- What new features would produce significant competitive advantage?
- What related business could be entered successfully and how?
- If the firm were to redesign its construction methods from scratch, what techniques would be adopted?

Short-range goals concentrate primarily on making the most out of the present-day situation. Scheduling, progress reporting, subcontracting, site supervision, estimating, purchasing, etc., are factors in determining and meeting short-range goals.

Scheduling

After planning decisions have been formalized, the schedule is established. Scheduling is totally time oriented. Time requirements for every operation depend upon the methods, materials, and manpower previously selected. The scheduler can compress, overlap, or distribute times, but he has very limited opportunities to change total manhours or leadtimes.

Good scheduling provides important advantages, many of them affecting overall project costs. Good scheduling:
- Reduces in-process time and minimizes time-related costs, that is, construction loan interest, insurance, security, overhead per unit, maintenance, and rate of capital utilization.
- Helps to smooth out manpower peaks and valleys and provides a more uniform flow of manpower requirements.
- Coordinates manpower and the other resource inputs. This effects a close integration of work in process. It insures that labor, materials, and subcontractors are brought together in a timely manner, and it furnishes a chronological checklist.
- Is a system for controlling work in process. The schedule offers all levels of personnel (management, production supervision, workman and subcontractors) a measurement of work accomplishment against target goals. A psychological incentive is provided for meeting the schedule objectives. The resulting feeling of accomplishment when staying even or ahead of schedule, contributes a positive benefit to everyone.
- Helps to identify potential problem areas.
- Assists in meeting delivery promises.

In addition, improved customer relations usually result because realistic completion dates can be met with a good scheduling system.

The first step in scheduling is to analyze the principal phases of building. Each operation is estimated for quantity of work, time span, sequence of installation, and operational interrelationships.

Sequential precedence is established by asking three questions about each operation:
1. What operations must or should be completed before this one can be started?
2. What operations can be worked concurrently with this one?
3. What operations must or should wait until this job is finished?

On the basis of these answers, critical operations are laid out in time sequence in accordance with the "must" answers.

Next, the "should" precedences are fitted in jigsaw-wise where they best benefit the schedule. Here the competent scheduler applies his knowledge of interrelations, costs, and skills to select the best choices from alternative sequences.

The determination is made in the office whether to build on a 40-day or a 120-day schedule. The project's craft noninterference patterns, provision for emergencies and material-due dates are established at this time. The success of a schedule depends primarily upon the scheduler's expertise at providing for a multitude of details, planned and unforeseen.

The scheduler must have his fingertips on a mass of basic data such as the following:

- Management policy on tight or loose schedules.
- The organization's historical capability (not a determining factor but one to be considered).
- Tooling capabilities.
- Crew size and availability.
- Material lead times.
- Costs of specific installations; postponing installation of expensive items on subcontracts, until necessary, reduces carrying charges.
- Workable days anticipated (holidays, weather problems, vacations).
- Cost of construction funding.
- Quality of supervision.
- Workman utilization data, to keep subs as well as employees on a smooth schedule.

The schedule policy established by management should be a flexible one. A policy such as "build from foundation to punch-list in 40 days" is neither flexible nor logical. A nicely rounded figure of 20 days or 60 days is the first hint of a directed (therefore, not logically developed) schedule. Statistically, only 10 percent of schedules should end in a nice round zero. In practice, however, a large number of 30, 60, or 90-day schedules are found. The schedule elapsed time should be a fallout of the scheduling process, not a directed goal.

Schedule spans are not directly comparable between builders or even between jobs. The variables of crews, availability of materials, weather, and scope of work contribute toward making each job a new set of conditions. Attempting to compare schedules requires considering questions such as:

- How much work was actually done within the schedule limits? Did the schedule include excavation to final punch-list?
- How much work was prefabricated offsite *before* the schedule initial date? (Work done offsite is no less a part of the schedule.)
- What percentage of units actually met the established schedule? (A schedule that is seldom met becomes, in effect, only a vague target.)

A too-tight schedule demands excessive supervisory control. It also leads to slippage which may require expensive cures. It does not contain sufficient slack for emergencies.

A too-loose schedule requires more costly construction financing. Its excess slack brings "Parkinson's Law" into effect resulting in low incentive and wasted time. Simply stated, Parkinson's Law means that work tends to expand to fit the time allotted. This in turn can lead to aimless organization and effort of work crews.

Everyone talks about inventory costs, but few have really tried to determine what these costs are. The scheduler should know the cost of an extra day, so that he has a rational basis for shortening or lengthening the schedule. Items to be considered when determining construction inventory costs include:

- Interest costs of construction funds.
- Site financing costs.
- Overhead items such as marketing, supervision, security, and the like.

The scheduler should weigh the costs of carrying completed units in finished inventory vs. the costs of a slower schedule that might obligate construction funds more slowly. In times of reduced sales activity, it may be beneficial to slow down rather than hurry up and wait for a buyer. On the other hand, if sales are brisk, a fast schedule reduces construction funding costs and also increases the turnover rate thereby improving the profit picture.

Material and other requirements should be scheduled the same as labor and subcontractors. Such items as permits, drawings, and tools could delay a project just as surely as a shortage of workmen.

In small building operations, the scheduling function is often accomplished by the owner. In larger companies, however, this task can be delegated to a subordinate who is in a position to work closely with the top echelon of management, estimators, and construction supervisors. Regardless of the size operation, experience shows that site superintendents should be involved in the planning of construction and scheduling. In addition, subcontractors should be consulted in preparing schedules. They should never be left to schedule their own work. The builder may have to accommodate to a subcontractor's availability, but this too should be coordinated by the builder's staff.

The optimum schedule is one that is reasonably attainable with some effort. It provides for unforeseen contingencies in the field. It can be used without difficulty by field supervisors and provides monitoring capability for management. Highly sophisticated scheduling systems are advantageous for large construction projects, but for small-to-medium home building enterprises, less complicated ones are satisfactory. The complexity of the scheduling system should not exceed that required for adequate control.

The Total Time To Complete Any Work Is Composed of Productive Times and the Nonproductive Times Between

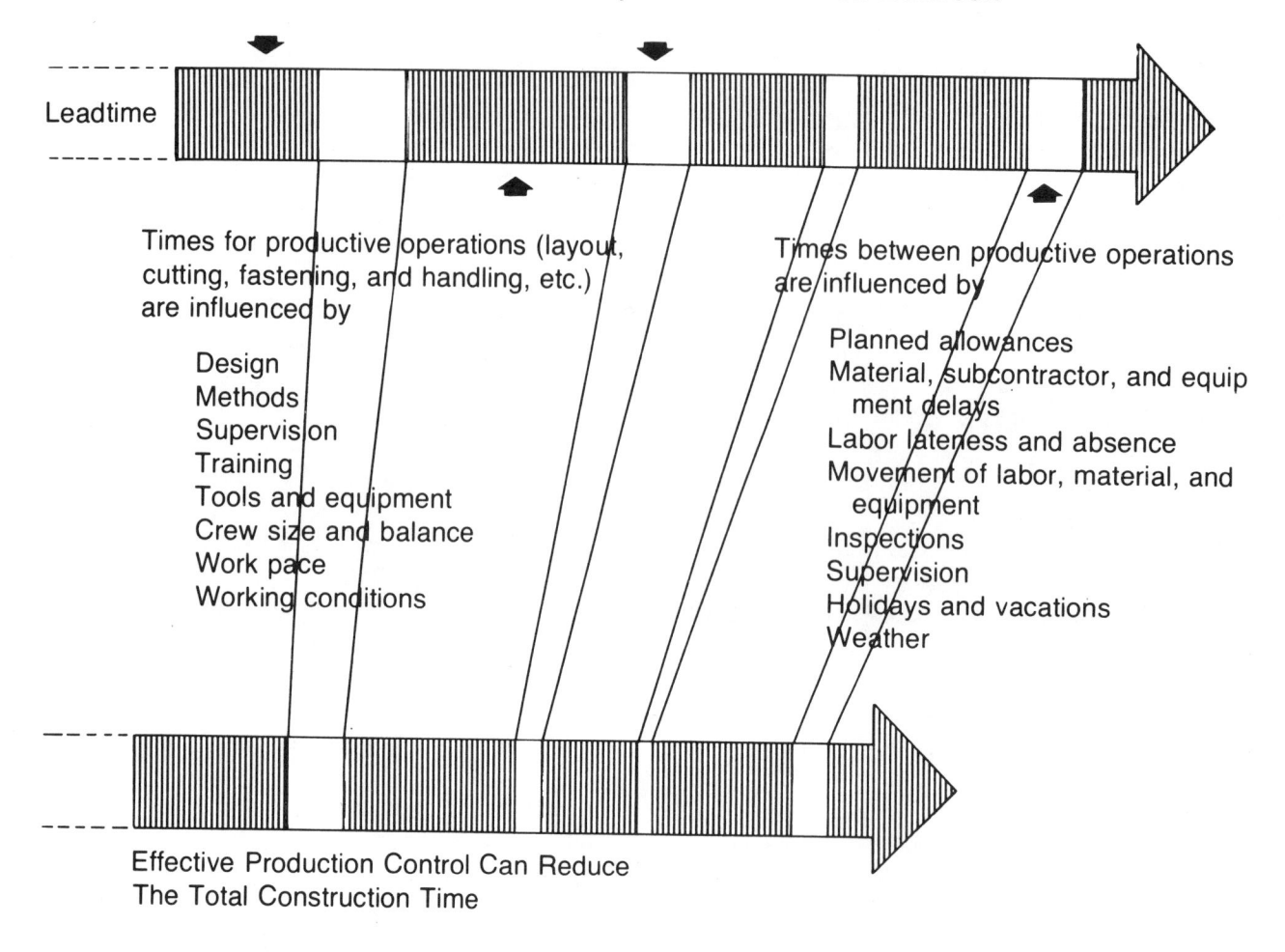

Leadtime

Times for productive operations (layout, cutting, fastening, and handling, etc.) are influenced by

Design
Methods
Supervision
Training
Tools and equipment
Crew size and balance
Work pace
Working conditions

Times between productive operations are influenced by

Planned allowances
Material, subcontractor, and equipment delays
Labor lateness and absence
Movement of labor, material, and equipment
Inspections
Supervision
Holidays and vacations
Weather

Effective Production Control Can Reduce
The Total Construction Time

The actual format of an effective schedule is relatively unimportant as long as the primary objective is achieved. That objective is to ensure that a planned and measured volume of construction activities are accomplished daily. A survey of many building operations might prove that no two scheduling systems are identical in all respects. They may range from sophisticated systems such as the Critical Path Method (CPM) to simple systems consisting of handwritten listings of work to be accomplished daily.

The degree of sophistication necessary to schedule properly depends on many factors, the major factor being the effectiveness of site supervision. A builder in Florida had a very complete and somewhat sophisticated wall chart schedule for one of several construction projects. Another similar project was being built by the same builder without a wall chart schedule. Both projects were well managed and both were being completed on time. When an analysis of the two projects was made, the second site was being managed by a very experienced superintendent who had the ability to control the job and "schedule" work without charts and graphs. The first site, equally well managed, had a superintendent with less experience who was unable to keep work flowing without a checklist or master schedule.

Since superintendents who are able to control entire projects without written schedules are rare, most builders would probably benefit from written schedules of some sort. Again, working schedules may be quite simple. *Sample Schedule for Garden Type Apartments* was used by a prominent Texas builder, and it can be seen to be nothing more than a listing of work operations and workdays. Much thought and experience went into making the schedule simple and easy to follow.

Sample Schedule for Garden Type Apartments*

Day	Operation No.	Operation
1	1	Layout first floor
	2	Stock first floor
2	1	Raise first floor walls
	2	Plumber stack out first floor
	3	Lay out floor joists
3	1	Floor joists
	2	First floor conduit
	3	Install vent ducts
	4	Stock subfloor
	5	Clean-up
4	1	Subfloor
	2	Deliver trusses
5	1	Layout second floor
	2	Stock second floor walls
	3	Set stairs
	4	Clean-up
6	1	Raise second floor walls
	2	Stock trusses
7	1	Raise trusses
	2	Plumber stack out second floor
	3	Stock outside doors
8	1	Roof lathe
	2	Second floor vents
	3	AC pipes
	4	Set up scaffolds
	5	Clean-up
9	1	Facia and soffit
	2	Second floor conduit
	3	Prime gutters
10	1	Shingle
	2	Siding
	3	Set outside doors
	4	Sheetrock furr down
	5	Tape furr down
11	1	AC duct
	2	Set AC unit
	3	Insulate AC pipes
	4	Hang gutters
	5	Install outside stairs

Day	Operation No.	Operation
12	1	Finish furr downs
	2	Exterior paint
	3	Clean-up inside
13	1	Pour lightweight concrete
14	1	Dry and inspect lightweight concrete
15	1	Plumber set tubs
	2	Electrician pull wires
	3	Insulate walls and ceiling
	4	Check inside framing
	5	Clean-up
16	1	Second inspection
	2	Stock sheetrock and soundboard
17	1	Soundboard
	2	Sheetrock
18	1	Tape
	2	Brick first level
19	1	Bed
	2	Brick
20	1	Bed
	2	Brick
21	1	Drybed
	2	Clean brick
22	1	Texture
	2	Grade for flat
23	1	Stock inside trim
	2	Form for flat
24	1	Inside trim
	2	Pour flat
25	1	Sand and Putty
26	1	Flat wall
27	1	Enamel
28	1	Stock cabinets
	2	Stock appliances
	3	Set closet doors
29	1	Set cabinets
	2	Set hoods

*Form from Jesse Harris Construction Co., Dallas, Texas.

Sample Schedule for Garden Type Apartments*

Day	Operation No.	Operation
30	1	Sand floors
31	1	Lay floor tile
	2	Finish outside paint
32	1	Wall tile
	2	Install fence
33	1	Plumbing fixtures
	2	Electric fixtures
	3	AC grills
	4	Hardware and base

Day	Operation No.	Operation
34	1	Clean windows
	2	Inside clean-up
	3	Grade yards
	4	Weatherstrip
	5	Landscape
35	1	Touch-up
36	1	Carpet
37	1	Final clean-up
38	1	Final inspection

Sample Scheduling System

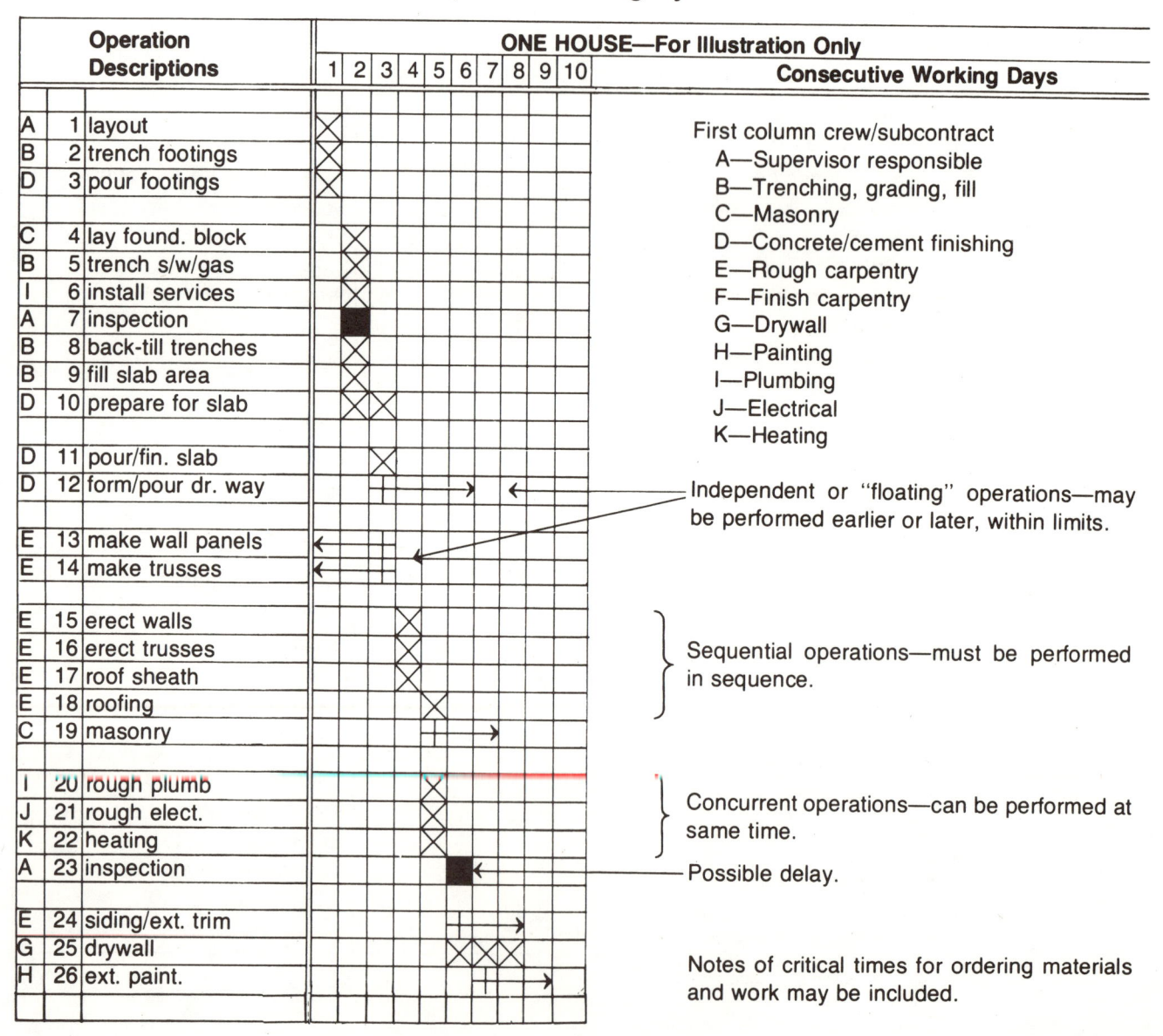

		Operation Descriptions	ONE HOUSE—For Illustration Only (Consecutive Working Days 1-10)
A	1	layout	
B	2	trench footings	
D	3	pour footings	
C	4	lay found. block	
B	5	trench s/w/gas	
I	6	install services	
A	7	inspection	
B	8	back-till trenches	
B	9	fill slab area	
D	10	prepare for slab	
D	11	pour/fin. slab	
D	12	form/pour dr. way	
E	13	make wall panels	
E	14	make trusses	
E	15	erect walls	
E	16	erect trusses	
E	17	roof sheath	
E	18	roofing	
C	19	masonry	
I	20	rough plumb	
J	21	rough elect.	
K	22	heating	
A	23	inspection	
E	24	siding/ext. trim	
G	25	drywall	
H	26	ext. paint.	

First column crew/subcontract
A—Supervisor responsible
B—Trenching, grading, fill
C—Masonry
D—Concrete/cement finishing
E—Rough carpentry
F—Finish carpentry
G—Drywall
H—Painting
I—Plumbing
J—Electrical
K—Heating

Independent or "floating" operations—may be performed earlier or later, within limits.

Sequential operations—must be performed in sequence.

Concurrent operations—can be performed at same time.

Possible delay.

Notes of critical times for ordering materials and work may be included.

Sample Weekly Field Schedule*

Location _____ Week Starting _____

| No. | Activity | Lot Numbers |||||| Comments |
		Mon.	Tues.	Weds.	Thur.	Fri.	Sat.	
1	Permits, Fees, Engineering							
2	Site Work							
3	Demolition							
4	Utility Connections							
5	Footings and Foundations							
6	Structural Steel							
7	Framing							
8	Concrete							
9	Rough Sheet Metal							
10	Rough Electrical							
11	Rough Plumbing							
12	Rough Heating, Vent., and A.C.							
13	Roofing							
14	Masonry							
15	Windows and Doors—Glazing							
16	Insulation							
17	Exterior Trim							
18	Exterior Paint							
19	Stairs							
20	Drywall and Plaster							
21	Ceramic Tile							
22	Finish Carpentry and Trim							
23	Flooring							
24	Cabinets and Vanities							
25	Interior Decoration							
26	Finish Electrical							
27	Finish Plumbing							
28	Finish Heating, Vent, and A.C.							
29	Finish Sheet Metal							
30	Appliances and Appointments							
31	Apartment Finish Items							
32	Building Clean-up							
33	Landscaping							
34	Final Inspection and Punch List Completion							

*Form from Place and Co., Inc., South Bend, Indiana.

Bar Chart Schedule

BAR CHART SCHEDULE	ACCT NO
I PREPARATION	
PERMITS, FEES, ENGINEERING	14301
SITE WORK	14302
DEMOLITION	14304
UTILITY CONNECTIONS	14305
FOOTINGS & FOUNDATIONS	14306
	14308
II ROUGH STRUCTURE	
STRUCTURAL STEEL	14310
FRAMING	14311
CONCRETE	14313
ROUGH SHEET METAL	14315
ROUGH ELECTRICAL	14317
ROUGH PLUMBING	14319
ROUGH HEATING VENTILATION & AIR CONDITIONING	14321
	14323
III FULL ENCLOSURE	
ROOFING	14330
MASONRY	14331
WINDOWS & DOORS—GLAZING	14333
INSULATION	14335
EXTERIOR TRIM	14337
EXTERIOR PAINT	14339
STAIRS	14341
	14345
IV FINISHING TRADES	
DRYWALL & PLASTER	14350
CERAMIC TILE	14351
FINISH CARPENTRY & TRIM	14353
FLOORING	14355
CABINETS & VANITIES	14358
INTERIOR DECORATION	14361
FINISH ELECTRICAL	14363
FINISH PLUMBING	14365
FINISH HEATING, VENTILATION & AIR CONDITIONING	14367
FINISH SHEET METAL	14369
APPLIANCES & APPOINTMENTS	14371
APARTMENT FINISH ITEMS	14373
	14375
V COMPLETION & INSPECTION	
BUILDING CLEAN UP	14380
LANDSCAPING	14381
FINAL INSPECTION & PUNCH LIST	14383
COMPLETION	14388
VI OTHER	14390

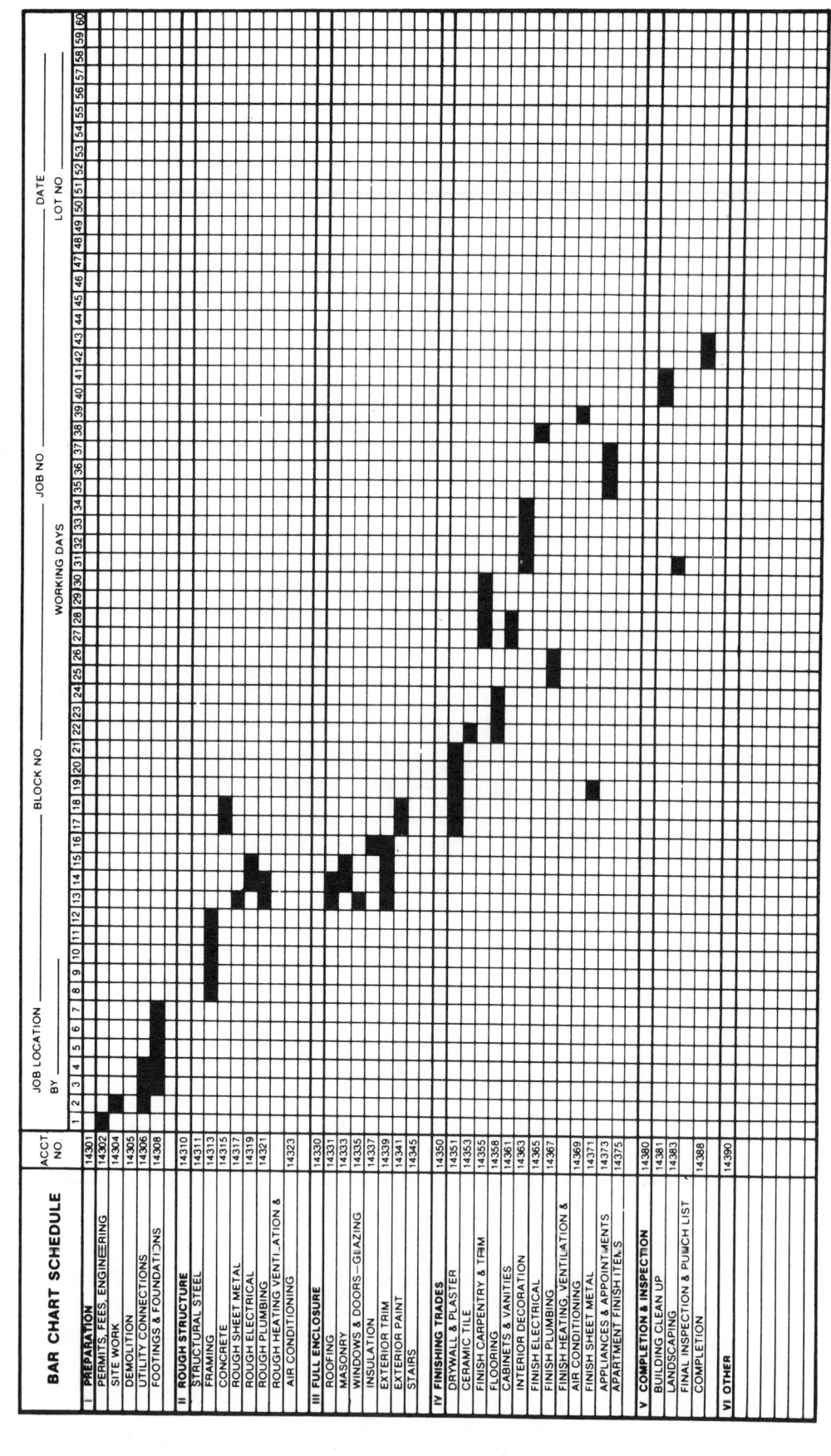

In *Sample Scheduling System*, some of the concepts of CPM are incorporated into a bar chart. Floating, sequential, and concurrent operations, all CPM terms, are accounted for in the scheduling system. Floating operations are somewhat independent of other operations and can be accomplished at any time, within limits. Operations marked with an X are considered critical and cannot be moved without disrupting the entire schedule. Solid squares are possible delay operations, that is, they are somewhat unpredictable. Operations scheduled with an arrow are floating operations and can be moved in the direction of the arrow within limits. Time standards for operations may be placed in squares rather than symbols.

Another builder uses a master schedule in his office but transmits details of the schedule to the field on a very simple form. If dates cannot be met for any reason, the site supervisors are required to inform the builder immediately so that his master schedule may be revised. *Sample Weekly Field Schedule* similar to that used by the builder.

Bar charts are often used for scheduling purposes. *Sample Bar Chart Schedule* shown on page 18 is based on *The Accounting System For All Builders* chart of accounts. This accounting system is used by many builder members of NAHB and provides a uniform chart of accounts for the industry. Bar charts such as the one shown can be made up in advance and issued when construction is started.[*] Because the schedule is made for working days instead of calendar days, schedule slippage for weather or other delays do not cause a schedule rewrite. A day delay is not considered a working day on this schedule.

Sample Chronological Schedule contains a sequence of construction operations, date scheduled, and date actually completed. A "by" column designates the crew or subcontractor responsible for performing each task. A thermometer bar, colored in as work is completed, flags problem areas and missed operations, offering a quick visual status check.

Sample Chronological Schedule

Step No.	Operation	By	Lot No. 705 Sched.	Actual	By	Lot No. 431 Sched.	Actual	By	Lot No. 432 Sched.	Actual	By
1.	Stakeout	P	4/26	4/28	P	5/2	5/2	D	5/4	5/4	D
2.	Excavate	C	4/29	4/30	C	5/3	5/3	C	5/7		C
3.	Footings	T	5/1	5/1	T	5/7		T	5/9		T
4.	Foundations	T	5/3	5/4	T	5/9		T	5/14		
5.	Waterproof	C	5/7		C	5/12		C	5/17		
6.											
7.	First inspection	A	5/10		Y	5/13		A	5/17		
8.	Backfill and rough grade	C	5/11		C	5/13		C	5/17		

Because the objective of a scheduling system is to ensure that a certain volume of work is accomplished daily, many builders insist on an inspection system to determine whether an activity was actually completed and if the quality was proper. Site supervisors cannot possibly remember every daily item or inspection affected at the job site. For this reason, daily records of inspections are often used. *Sample Heating and Air Conditioning Checklist* was used by a Texas builder; he has summary checklists for each construction activity.

---

[*](1987 ed.) NAHB now recommends the chart of accounts in *Accounting and Financial Management*, National Association of Home Builders, Washington, D.C.

Sample Heating and Air Conditioning Checklist*

Apt. No. _____ Bldg. No. _____ Inspected by _____ Date _____

Building Exterior
- ____ Valve box cover in place
- ____ Valve box cover to grade
- ____ Valve box cover aligned

First Floor Heating/Cooling
- ____ Thermostat/speed switch aligned
- ____ Polarity correct
- ____ Speed switch operates
- ____ Fan quiet
- ____ Hi/low speed
- ____ Dining room supply grill set
- ____ Dining room supply grill not damaged
- ____ Kitchen supply grill set properly
- ____ Kitchen damper knob attached
- ____ Kitchen supply grill not damaged
- ____ Return grill set properly
- ____ Return grill not damaged
- ____ Filter in place
- ____ Grill nuts set
- ____ Grill opens easily
- ____ Access door not damaged
- ____ Access door painted
- ____ Access door fit
- ____ Access door not warped
- ____ Access door clear
- ____ Access door easy action
- ____ Access door hardware installed properly
- ____ Access door hardware not damaged
- ____ Plenum holes sealed
- ____ Fans and shafts operate smoothly
- ____ Fan motors not dripping oil
- ____ Resistance heater operates
- ____ Condensate not leaking
- ____ Disconnect operates
- ____ Disconnect box sufficient size
- ____ Disconnect box attached

Second Floor Heating/Cooling
- ____ Thermostat/speed switch at jamb
- ____ Thermostat/speed switch aligned
- ____ Polarity correct
- ____ Speed switch operates
- ____ Fan quiet
- ____ Hi/low speed
- ____ Bedroom No. 1 (front) supply grill set properly
- ____ Bedroom No. 1 supply grill not damaged
- ____ Bath No. 1 supply grill set properly
- ____ Bath No. 1 supply grill not damaged
- ____ Bath No. 2 supply grill set properly
- ____ Bath No. 2 supply grill not damaged
- ____ Bath No. 2 damper knob attached
- ____ Door clears damper knob
- ____ Bedroom No. 2 (rear) supply grill set properly
- ____ Bedroom No. 2 supply grill not damaged
- ____ Return grill set properly
- ____ Return grill not damaged
- ____ Return air filter in place
- ____ Return air grills nuts set
- ____ Return air grill opens easily
- ____ HAC door bit damaged
- ____ HAC door painted
- ____ HAC door fit
- ____ HAC door not warped
- ____ HAC door clean
- ____ HAC door easy action
- ____ HAC door hardware properly installed
- ____ HAC door hardware not damaged
- ____ Plenum holes sealed
- ____ Fan and shafts operate smoothly
- ____ Fan motors not dripping oil
- ____ Resistance heater operates
- ____ Condensate not leaking
- ____ Disconnect operates
- ____ Disconnect box sufficient size
- ____ Disconnect box attached
- ____ Visible insulation installed properly
- ____ Visible insulation not damaged

Another inspection sheet format, *Sample Construction Record for Unit and Inspection*, can be used occasionally as a check on the actual construction record for the purpose of adjusting the schedule if necessary. As the Inspector (usually the superintendent) inspects each phase of construction, he enters the date that the work was completed in the square next to the work description. For example "6/8 Footings" simply means footing work was completed June 8. Depending on the characteristics of the dwelling and construction operation, boxes and subjects may be added or eliminated from this page. This then is used to update the schedule or revise subsequent schedules according to performance.

---

*Form from Jesse Harris Construction Co., Dallas, Texas.

Sample Construction Record for Unit and Inspection

Plat _____ Section _____ Lot _____ Sheet ___ 1 of 4 ___

Phase I

☐ **Stage A—Site**
- ☐ Legal
- ☐ Services
- ☐ Access
- ☐ Siting

- ☐ Record
- ☐ Electricity Telephone
- ☐ Street
- ☐ Elevation

- ☐ Survey
- ☐ Gas Water
- ☐ Driveway
- ☐ Setbacks

- ☐ Permits
- ☐ Sewage
- ☐ Walk
- ☐ Stake House

- ☐
- ☐
- ☐
- ☐ Stake Drive

☐ **Stage B—Earthwork**
- ☐ Excavate
- ☐ Trench
- ☐ Cut
- ☐ Cleanup Haul

- ☐ Foundation Areaway
- ☐ Garage
- ☐ Driveway
- ☐ Debris

- ☐ Spot Top Soil
- ☐ Piping Wiring
- ☐ Spot Top Soil
- ☐ Excess Fill

- ☐ Spot Back Fill
- ☐ Inspect
- ☐ Spot Back Fill
- ☐

- ☐ Spot Fill Excess
- ☐ Back Fill
- ☐ Spot Fill Excess
- ☐

☐ **Stage C—Foundation**
- ☐ Footings
- ☐ Walls
- ☐ Walls
- ☐ Other

- ☐ Forms reinf.
- ☐ Courses level-sq.
- ☐ Basement Garage
- ☐ Waterproofing

- ☐ Pour
- ☐ Pilasters Piers
- ☐ Areaway Fireplace
- ☐ Window Wells

- ☐ Strip
- ☐ Jambs Lintels
- ☐ Door/Win. Beam/Col.
- ☐

- ☐ Repair
- ☐ Wall anchorage
- ☐
- ☐ Inspect

☐ **Stage D—Miscellaneous**
- ☐ Rough Mechanical Pipe and Wire
- ☐ Gravel Fill
- ☐ Salvage
- ☐ Cleanup Haul

- ☐ Ground P and W
- ☐ Basement Garage floors
- ☐ Form
- ☐ Excess Material

- ☐ P and W Inspect
- ☐ Footings Drains
- ☐ Scaffold
- ☐ Rubbish

- ☐ Temporary Electric
- ☐ Garage Floor
- ☐ Masonry
- ☐

- ☐
- ☐ Driveway
- ☐ Sand Gravel
- ☐ Inspect

Sample Construction Record for Unit and Inspection (Cont.)

Plat _____ Section _____ Lot _____ Sheet ___2 of 4___

Phase II

☐ **Stage A—Floors**

☐ Framing	☐ Sill Plate Level-Sq.	☐ Insulate	☐ Joist, Grade Spacing	☐ Fasten
☐ Openings	☐ Stairs	☐ Fireplace	☐ Laundry Chute	☐ Service Access
☐ Subflooring	☐ Spacing	☐ Fasten	☐ Cutouts	☐
☐ Other	☐ Basement Stairs	☐	☐ Salvage Material	☐ Inspect

☐ **Stage B—Walls and Partitions**

☐ Framing	☐ Plumb Level	☐ Plates Studs	☐ Rough Openings	☐ Fasten
☐ Framing	☐	☐ Sheathing Siding	☐ Opening Trim	☐ Exterior Trim
☐ Framing	☐ Plumb Level	☐ Plates Studs	☐ Rough Openings	☐ Fasten
☐ Miscellaneous	☐	☐ Bulkhead Stair Rake	☐ Backers Nailers	☐ Inspect

☐ **Stage C—Roof**

☐ Framing	☐ Truss Spacing	☐ Ties, Blocking	☐ Fasten	☐ Catwalk
☐ Framing	☐ Gable Ends Louvers	☐ Gable End Sheathing	☐ Soffits Rakes	☐ Exterior Trim
☐ Sheathing	☐ Spacing Clips	☐ Cutouts	☐ Fasten	☐ Inspect
☐ Shingles	☐ Metal Flashing	☐ Starter Ridge	☐ Fasten	☐ Inspect

☐ **Stage D—Miscellaneous**

☐ Other	☐ Repairs	☐ Cut Misc. Openings	☐ Exterior Doors	☐ Gutter System
☐ Salvage	☐ Lumber Plywood	☐ Siding Trim	☐ Metal Fasteners	☐ Shingles
☐ Fill-Grade	☐ House Garage	☐ Spread Top Soil	☐ Gravel Fill Walk/Drive	☐ Spot Excess
☐ Cleanup Haul	☐ Excess Fill	☐ Salvage Material	☐ Rubbish	☐ Gravel Bank Inspect

Sample Construction Record for Unit and Inspection (Cont.)

Plat _____ Section _____ Lot _____ Sheet __3 of 4__

Phase III

☐ **Stage A—Rough Mechanicals**

☐ Electricity Telephone	☐ Ent. Panel Circuits	☐ Outlets Switches	☐ Thermostat	☐ Inspect
☐ HVAC	☐ Furnace/HWH Venting	☐ Ductwork Openings	☐ Air Fans	☐ Inspect
☐ Plumbing Gas	☐ Main Branches	☐ Outlets Controls	☐ Furnace/HWH Range/Dryer	☐ Inspect
☐ Water Sewage	☐ Main Branches	☐ Supply DWV	☐ Outlets Pop-Off	☐ Inspect

☐ **Stage B—Concrete, Masonry, and Metals**

☐ Concrete Forms, Reinf.	☐ Basement Floor Areaway	☐ Step/Walk AC Pad	☐ Garage Floor Apron	☐ Driveway
☐ Concrete: Pour, Strip	☐ Basement Floor	☐ Step/Walk AC Pad	☐ Garage Floor	☐ Driveway
☐ Masonry	☐ Fireplace Chimney	☐ Planter	☐	☐
☐ Metals	☐ Chimney Flashing	☐ Areaway Railing	☐	☐

☐ **Stage C—Insulation and Drywall**

☐ Insulate	☐ House/Garage Ceilings	☐ House/Garage Ext. Walls	☐ Salvage Cleanup	☐ Inspect
☐ Gypsumboard	☐ House/Garage Ceilings	☐ House/Garage Walls	☐ Bulkhead Stairwell	☐ Closets
☐ Joints	☐ House/Garage Taped	☐ House/Garage Finished	☐	☐ Inspect
☐ Miscellaneous	☐ Salvage	☐ Cleanup, Haul	☐	☐

☐ **Stage D—Trim and Paint** (Note: All interior trim and doors painted)

☐ Flooring	☐ Underlay Wood Strip	☐ Bath Lavatory	☐ Kitchen	☐ Inspect
☐ Walls	☐ Wood Panel	☐ Bath	☐	☐ Inspect
☐ Trim	☐ Closets Window Trim	☐ Baseboard Stair rail	☐ Ext. Doors Misc.	☐ Inspect
☐ Paint/Stain	☐ Ceiling Wall Prime	☐ Ceiling Wall Finish	☐ Sand/Finish Wood Floors	☐ Inspect

Sample Construction Record for Unit and Inspection (Cont.)

Plat _____ Section _____ Lot _____ Sheet __4 of 4__

Phase IV

☐ **Stage A—Fixtures and Fittings**
☐ Cabinetry Tops	☐ Basement Wall Kitchen	☐ Vanity Bath	☐	☐ Inspect
☐ Electricity Telephone TV	☐ Ceiling/Wall Lights	☐ Switches Outlets	☐ Spec. Purp. Misc.	☐ Inspect
☐ Heat/AC	☐ Sup. Return Registers	☐ AC Coil/ Condenser	☐	☐ Inspect
☐ Plumbing	☐ Bath Lavatory	☐ Kitchen Laundry	☐	☐ Inspect

☐ **Stage B—Appliances and Accessories**
☐ Appliances	☐ Refrigerator Range	☐ Washer, Dryer	☐	☐ Inspect
☐ Accessories	☐ Kitchen	☐ Laundry	☐ Bath Lavatory	☐ Inspect
☐ Vents-Chute	☐ Range Exhaust Fan	☐ Bath/Lav. Exhaust Fan	☐ Dryer Vent Clothes Chute	☐ Inspect

☐ **Stage C—Hardware and Carpet**
☐ Door	☐ Int./Ext. Locksets	☐ Comb. Doors Stops	☐ Weatherstrip Thresholds	☐
☐ Window	☐ Operators Locksets	☐ Screens, Storms	☐ Blinds, Shades	☐
☐ Other	☐ Rods Hooks	☐ Pegboard, Garage	☐	☐
☐ Carpet and Pad	☐ Shoe Edging	☐	☐	☐ Inspect

☐ **Stage D—Miscellaneous**
☐ Painting	☐ Basement Washer/Dryer	☐ Beams	☐ Stairs	☐
☐ Cleaning	☐ Fixtures Fittings	☐ Appliance Access	☐ Glass Flooring	☐ Basement Garage
☐ Cleanup Haul	☐ Salvage Materials	☐ Rubbish	☐	☐ Inspect
☐ Inspection	☐ Builder	☐ Buyer	☐ Bank	☐

More elaborate line-of-balance charts (for high production) or the Critical Path Method (for complex projects) may be useful to some builders. These systems should be investigated as scheduling alternatives. As previously mentioned, any scheduling system is only as good as the personnel who operate it. One builder who installed CPM was convinced that the system was responsible for his excellent record of remaining on schedule. Years later, when his superintendent admitted that he never understood CPM and was running the job without ever referring to the schedule, the builder stopped praising the CPM scheduling system and started exhorting the value of good supervision.

Supervision

Proper site management is fundamental to a successful home building operation. Proper site management means "absolute" control of one's resources including materials, labor, and equipment. The management control function starts with the highest echelon, and from that point, it is delegated through the organizational chain to first-line supervisors. Management, like a chain, is only as strong as its weakest link. Top management is only as strong or as effective as site supervision, and likewise, site supervision is only as strong or as effective as top management allows it to be.

The backbone of production is the onsite supervisory personnel. Regardless of the building volume, good site supervision usually results in quality construction, satisfied customers, reasonable profits, and the continuing good reputation for the builder. The site supervisor has three major responsibilities:
- To get out production.
- To hold down costs.
- To train, help, guide, and lead the production people.

The effective site supervisor needs three major skill requirements:
- Technical ability to use knowledge, methods, techniques, and equipment necessary for the performance of specific tasks.
- Human ability and judgment in working with and through people, including the understanding of motivation and application of effective leadership.
- Conceptual ability to understand the complexities of the overall builder's organization and where his operation fits into the overall organization.

Conceptual ability permits him to act according to the objectives of the total organization rather than only on the basis of his own group's goals and needs. Conceptual skills are very important to cost-effective construction systems.

An effective site superintendent, among other things, will:
- Insist on high standards.
- Be on top of all things under his jurisdiction at all times.
- Challenge his employees to do a job of which they can be proud.
- Search for ways to improve the operation.
- Give clear, timely instructions.
- Coordinate all building operations.
- Develop systematic means of knowing how the work is progressing.
- Be interested in and considerate of each employee.
- Help and encourage his employees to develop their activities.
- Be a good listener.
- Permit employees to participate in the decision making process.

The success of site supervision ultimately rests upon the superintendent's ability to build a strong, stable team. That means the ability to attract the best men, to recognize their potential and build upon it. The successful building organization usually has one major attribute that sets it apart from unsuccessful building organizations: dynamic and effective leadership. To achieve leadership, however, requires a complete understanding between the top echelon of management and field supervision. The two elements must have a definite understanding of authority, responsibility, and accountability.

Authority must certainly be commensurate with responsibility for the superintendent to be effective at the jobsite. The amount of authority delegated to the superintendent, however, must never exceed the amount earned by that superintendent. One of the definitions of authority given by Webster is "the power to influence . . . behavior." Authority also has four lesser ingredients, each very much a part of effective leadership. These are: position, competence, personality, and character.

One in authority is chosen for the position. Those reporting to him are supposed to do what he tells them to. His competence has been learned from experience and education, and his crew members should be confident that his decisions are sound. He has developed a personality that determines the degree of ease of doing business and working with him and talking and listening to him. The character of the person in authority is perceived by others. It is not what he believes it is, but what others believe it to be.

Organization

An organization translates plans into action. An organizational chart shows each person his place in the organization. On every project, one person should be responsible for onsite construction for the entire project. This person is called the site superintendent in this manual. In cases where more than one construction site is being developed, a general superintendent may be responsible for overall control of all building sites. Depending upon building volume, phase superintendents also may be assigned. Each phase superintendent is responsible for his phase of construction scheduled for that week. Phase superintendents report directly to the site superintendent. The site superintendent and, usually, phase supervisors are nonworking supervisory personnel (that is, they perform no direct labor). NAHB Research Foundation field studies show that the supervisor is much more valuable when he is supervising than when he is doing work. However, foremen (leadmen or crew leaders) in charge of small groups of workers perform work along with their men. Subcontractors in many organizations may function as foremen. In any case, they should be responsible directly to the site superintendent or his designate for the respective construction work. Starting with the foreman, it is good practice for each supervisor to be paid more (10 to 15 percent) than the highest paid employee reporting to him. Otherwise, an employee has no incentive to be an effective supervisor or to assume supervisory responsibilities. Obviously, some building companies require more elaborate organizational structures and some can have simpler organizations than the one illustrated. Nevertheless, the basic principles are applicable to all building firms, large and small. Note: In many organizations, subcontractors will function in the capacity of foremen. They should be organized in the same manner as outlined in *Sample Construction Organization Chart*.

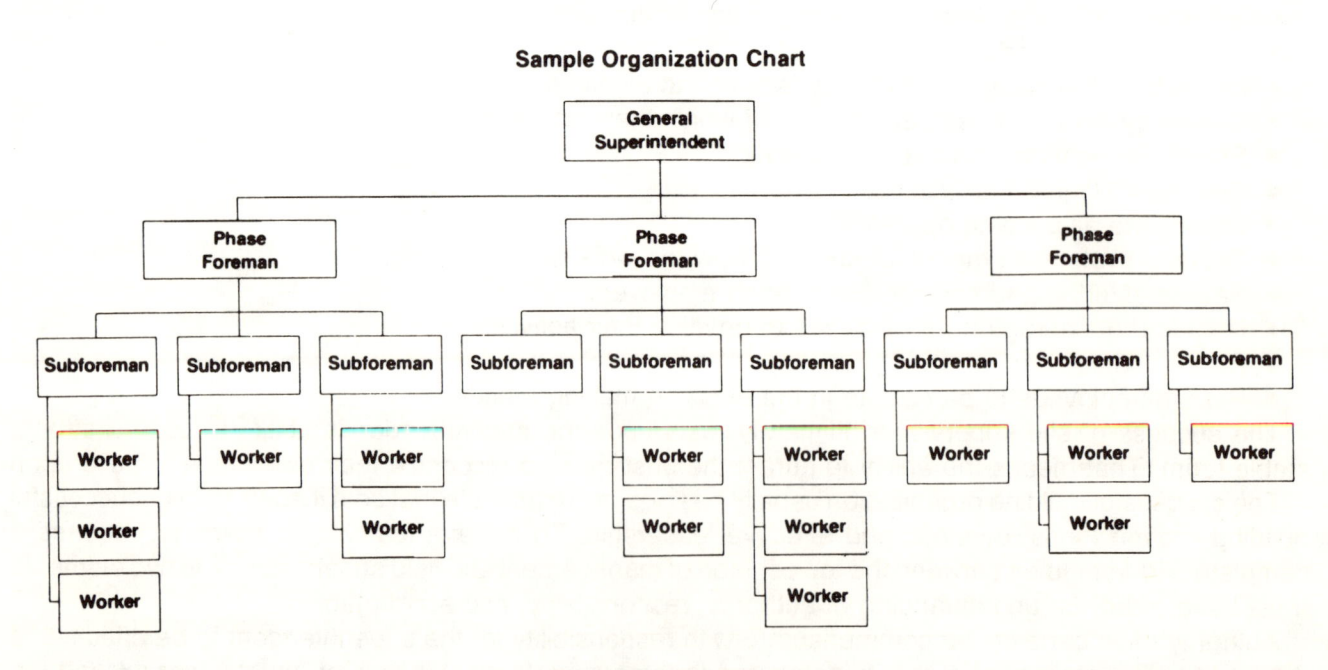

Sample Organization Chart

NOTE: In many organizations subcontractors will function in the capacity of subforemen. They should be organized in the same manner as outlined above.

Job Description

Sample Job Description Checklist shows only the skeleton framework around which an organization is developed. The authority and responsibility at all supervisory levels must be defined by individual job descriptions. Once job descriptions have been developed, all employees of the organization should be made aware of the job description contents.

Sample Job Description Sheet	
Job Title:	General superintendent, phase foreman, subforeman, etc.
Department:	Construction
Date:	(Important in case of revisions.)
Statement of Job:	(A sentence or paragraph describing in general terms the functions and responsibilities of this particular job assignment.)
Duties:	(The duties of this job should be numbered and listed. If care is exercised in describing each job and in listing the duties of each job, the usual problems of overlapping responsibility and authority will be minimized or eliminated.)
Experience: Education:	(The amount of experience and education required to qualify for each job should be carefully spelled out. In some cases, experience may compensate for formal educational requirements.)

Span of Control

Span of control is defined as the number of tradesmen or housing units under construction that one supervisor can supervise adequately. By applying the principles of span of control, the builder and his site superintendent can establish a plan whereby optimum site supervision is maintained for each building site. The principles of span of control follow:

• Each person should know to whom he reports and who reports to him. Subordinates and subcontractors should positively know the identity of the person responsible for the supervision of their work. Relationships may be clearly defined at the outset but often they can be upset by the actions of top executives in countermanding instructions directly without going through the proper channels. Failure to follow definite relationships once they have been established tends to undermine authority and results in needless confusion that can be extremely costly.

• No person at the construction site should report to more than one supervisor unless it is in a staff function, regardless of the size of the building operation.

• An excessive number of persons should not report directly to one supervisor. No flat rule exists that states the optimum number of people that one supervisor can supervise effectively. The guiding factor in determining the practical limit is the nature of the work and the extent to which it may be difficult to supervise. For example, the foreman who finds no difficulty in supervising ten laborers backfilling a foundation may find it impossible to control the work of five craftsmen finishing two widely separated housing units.

The supervisory requirements needed to control construction adequately can vary from company to company. However, certain factors and basic assumptions can be used to help determine supervisory requirements. For example, some type of construction activity is needed in each housing unit at least every working day. Since the unit should not stand idle, construction should be so that one task at least is performed daily. A visual inspection will show what has or has not occurred during the day's work. Thus, daily inspection by the site superintendent becomes a significant factor in determining work progress and quality. Daily inspections are influencing workload factors from which the number of supervisory positions can be determined.

A cost-effective building system places emphasis on the need for daily inspections to ensure that the full benefits of labor and material economies are incorporated in the construction process. *Housing Unit Inspections Per Available Hours*, is based on the assumption that about 0.3 manhours, or from 15 to 20 minutes, is needed, on the average, for each inspection. The amount and type of work done in the unit between inspections and the experience of the superintendent will dictate the actual time required to inspect each individual dwelling. If a superintendent budgets 4 hours per day for inspections, he should be able to inspect 13 dwelling units per day. *Housing Unit Inspections Per Available Hours* shows the number of unit inspections possible per superintendent according to the number of hours allocated to inspections.

Housing Unit Inspections Per Available Hours

Unit Inspections	Manhours Allocated to Unit Inspections (*Assuming 15 to 20 minutes Per Inspection*)							
	1	2	3	4	5	6	7	8
Number possible per superintendent	3	7	10	13	17	20	23	27

Average Number of Units in Construction Flow shows a breakdown by construction schedule time and unit completions per week.

It indicates, for example, that for a weekly completion rate of 2.5 dwelling units (125 units per year) and a 14-week construction schedule, an average of 35 units will be in the construction flow at any one time. In addition, it shows the effect of reduced schedule time on supervisory requirements. At a completion rate of 2.5 dwellings per week and a 10-week construction schedule (instead of 14 weeks), an average of 25 units will be in the construction flow, reducing daily inspection requirements by one-third.

Average Number of Units in Construction Flow

Units Completions Per Week	Scheduled Completion Time (*In Weeks*)						
	4	6	8	10	12	14	16
1.0	4	6	8	10	12	14	16
1.2	6	9	12	15	18	21	24
2.0	8	12	16	20	24	28	32
2.5	10	15	20	25	30	35	40
3.0	12	18	24	30	36	42	48
3.5	14	21	28	35	42	49	56
4.0	16	24	32	40	48	56	64
4.5	18	27	36	45	54	63	72
5.0	20	30	40	50	60	70	80

Number of Superintendents Required shows how many superintendents are needed in relation to *Average Number of Units in Construction Flow* and according to *Housing Unit Inspections Per Available Hours*. For example, if 4 hours per day are allotted to inspections, it shows that 3 superintendents are needed if 35 units are in the construction flow.

Number of Superintendents Required
(Assuming 15 to 20 Minutes Per Inspection)

Number of Units In Construction Flow	Hours Per Day Allotted to Inspections				
	4	5	6	7	8
15	1	1	1	1	1
20	2	2	1	1	1
25	2	2	2	1	1
30	3	2	2	2	1
35	3	2	2	2	2
40	3	3	2	2	2
50	4	3	3	3	2
60	5	4	3	3	2
70	6	4	4	3	3
80	6	5	4	4	3

Subcontracting

Subcontracts account for the majority of craft work in home building. Nationwide, subcontractors account for:
- 68 percent of carpentry
- 88 percent of concrete flatwork
- 90 percent of roofing
- 94 percent of masonry
- 96 percent of drywall/plaster
- 96 percent of plumbing
- 96 percent of electrical wiring
- 97 percent of heating/cooling

This high incidence of subcontracting results from recognizing the advantages of the subcontracting system. The advantages are these:
- Specialization tends to improve productivity.
- Less builder supervision is required for subcontractor day-to-day work.
- Builders' supervisors do not need specialized tool skills but should have craft knowledge.
- Price is known in advance.
- Builder is not required to hire full-time specialized skills.
- Equipment inventory and maintenance are not required.
- Union negotiations may be handled by the subcontractor.
- Subcontractors are often low-overhead organizations.

These advantages are coupled with these disadvantages:
- Poor management may be found in subcontractors' organizations.
- Competition may arise among builders bidding for subcontractors, when the supply of competent ones is inadequate.
- The quality of workmanship and personnel policies may be hard to control.
- The possibility exists that subcontractors may be underfinanced and subject to "going broke."

The two primary principles of subcontracting are:
- Negotiate prudently, keeping in mind though that subcontractors cannot perform properly or stay in business unless they make a reasonable profit.
- Have a clear-cut agreement in writing to define all tasks, conditions, who supplies what, and price.

Negotiation is a way of life in homebuilding. Even fixed-rate subcontracts may be open to negotiation. One builder recently turned a small portion of finish carpentry over to his own crew, reducing the subcontractor's scope of work. His subcontractor, however, was adamant that he would not reduce the per-door cost. Upon further negotiations, he agreed to keep the same per-door price but to charge for one less door per house than the actual count. In spite of his no-negotiation policy, the subcontractor was still willing to accommodate some bargaining.

The use of written specifications for subcontracted materials and labor can improve quality, and frequently, lower cost. There are several advantages to using specifications for subcontracting:
- The designer is required to decide on many details which, when left unspecified, can result in builder headaches later.
- There is better mutual understanding when the specifications are reduced to writing.
- Quality or warranty disputes are more readily clarified and satisfactorily settled.
- The subcontractor may perform spec-quality work at standard-quality price by averaging costs for all jobs.

A clear channel of communication and feedback should be established between the builder or his representative and the subcontractors. Eliciting subs' suggestions and complaints and recognizing their strengths and weaknesses is beneficial in the long run. Subcontractors provide a rich source of information and suggestions. Of course, the builder or his personnel need to know enough about craft productivity and quality to evaluate properly such suggestions and information.

Subcontractor Selection

When selecting a subcontractor, the builder has a responsibility to make an eyes-open decision. Since the perfect subcontractor is difficult to find and may be more difficult to hold, the builder may have to suboptimize—to take less than the best. Many factors can make a subcontractor less than perfect; such characteristics as high bids, underfinancing, poor management, ineffective equipment, unreliable scheduling, uncertain quality, and shoddy workmanship all contribute to subcontractor deficiencies. *Sample Subcontractor Evaluation Form* can be helpful in the selection process.

The form shows the scoring of four available subcontractors. Subcontractor A is an established, high quality firm but likely to be higher in cost. Subcontractor B is average in cost and adequate in performance. Subcontractor C is low in cost, adequate in performance, but may be financially shaky. Subcontractor D is average in cost, high in quality, but marginal in meeting schedules.

Which of these subcontractors would be the best selection? The answer—any of them might be. It depends on the building situation.

According to some textbooks, Subcontractor C is an unlikely choice since he lacks stability and financial backing. Yet, some builders, both large and small volume, encourage this type of subcontractor. They

Sample Subcontractor Evaluation Form

Quality	Sub A	Sub B	Sub C	Sub D
Competitive Cost	2	3	4	3
Workmanship	4	3	3	4
Dependability	4	1	3	3
Schedule	3	3	3	2
Finance and Credit	4	3	2	3
Stability	4	3	2	3
Total	21	16	17	18

4—Superior
3—Adequate
2—Marginal
1—Don't know

use the builder's own financial structure to support the subcontractor's underfinanced condition. They provide closer supervisory coverage to compensate for his lack of managerial proficiency. At the same time, they benefit from his low cost, adequate workmanship for the job requirements, and close working relationship. At the other extreme, a builder of expensive custom homes might find subcontractor A to be the best choice. Or if schedules are not critical, he may elect the lower cost of D. In any event, the essential need is to evaluate subcontractors to fit specific situations.

Subcontractor Bidding

Friendly relations with subcontractors are, of course, beneficial, but the sub should be a subcontractor first and a friend second. Negotiating exclusively with old-line subcontractors is a form of habit buying, often not a cost-effective practice. The expedient of requesting bids from several qualified subs does not always solve the problem of stirring up competition. To attract new subs, a sealed-bid method can be used, this guarantees to the new sub that his quotation will not be used to lower a favored subcontractor's bid.

In times of material shortages and escalating prices, an escalator clause may be frequently requested by subcontractors. This removes the builder's advantage of knowing the full price in advance. There is no really good way to solve this problem. However some builders agree to certain escalation factors tied to some index or agree to renegotiate prices at stated intervals, such as 3 or 6 months. Otherwise the sub's alternative on a long-term contract is to put a gross inflation factor into his bid which may well be excessive.

Writing Subcontracts

Sample subcontracts are replete with boiler-plate clauses spelling out the subcontractor's and builder's responsibilities. The effective subcontract should be prepared by an attorney familiar with local laws and regulations. This can be prepared for an individual builder or as a series of optional clauses for a group or association of builders. Types of clauses frequently found in subcontracts include the following:
- Scope of work and specifications.
- Schedule, time of the essence.
- Responsibility for delay in work.
- Cleanup.
- Failure to perform.
- Termination upon specified time (48 hours?) notice (kick-out clause).
- Entire agreement (no other implied or verbal agreement).
- Warranty.
- Liens responsibility.
- Payment responsibility for licenses, permits, and taxes.
- OSHA regulations.
- Liability for injury and damage.
- Terms of payment.
- No subletting of contract.
- No assignment of payment.
- Responsible for own materials and equipment.
- Subcontractor's examination of site conditions.
- Correction in writing of errors or any changes.
- Subcontractor to notify builder of any problems.
- Control of employees.

- Employer's liability or workman's compensation.
- Equal employment opportunity.
- Fair labor standards, wages and hours.
- Collective bargaining.
- Penalty clauses.

Subcontractor Supervision

In the typical company, day-by-day contact with subcontractors is maintained by the site supervisor. To be effective, the supervisor needs:
- A clear understanding of the builder's policies and the supervisor's authority under those policies.
- A detailed, but flexible, schedule.
- Sufficient craft and quality requirements knowledge to evaluate the work of each trade.
- Understanding of subcontractor's problems.
- A diplomatic disposition, firm manner and a tough hide.

The supervisor coordinates both with and among subcontractors. He is responsible for scheduling them into work areas ready for their tasks. He resolves disputes or interference between crafts. He serves as a communication channel in all directions. In short, he is the center of coordination.

Quality Control

With the rise in consumerism, the home builder has been feeling increasing pressure on quality. Quality has some savings potential by reducing callback costs. Builders say though that the value of good will and customer referrals resulting from attention to quality is more important.

Quality control and inspection functions are usually delegated to site supervisors, although organizationally, this is far from ideal. A supervisor is responsible for rapid and economical completion of his projects. Indeed, he may be paid a bonus based on those criteria. To assign him the additional duty of policing himself can lead to a conflict of interest. On the other hand, a separate inspection department can be an expensive function and not always effective. Inspectors (including building code inspectors) sometimes believe they have a mandate to nit-pick details, forgetting that their function is productive quality control.

Quality control is not a matter of higher cost or closer inspection; it is a matter of better habits. Its principal maxim is: "quality is built in, not inspected in."

Inspection is not quality control; it is one of the tools of quality control, a post mortem examination to determine what was done wrong. Rejected items are then subject to rework, an expensive operational sequence.

An effective quality control program instills quality habits in workers. Every craftsman has a standard of quality. The level of standard varies from individual to individual and from job to job. The secret of quality assurance is to replace the craftsman's uncertain level of quality with an adequate standard of quality. Once the level is established and understood, workers have less difficulty working on that accepted level.

When management establishes a quality control policy, it must set standards. These are in the form of written, definite, clearly understood standards of quality held by all levels of builder's and subcontractor's personnel. They must be enforced without fail. The craftsman allowed to get away with poor workmanship tends to lower his quality standards. In homebuilding, nearly all substandard material or shoddy workmanship is probably recognized by the craftsman making the installation. Quality control corrects it at that time. Inspection results in rework later.

An effective quality control program will reduce homeowner complaints, but it cannot stop them. The callback program should include adequate feedback information to quality control personnel.

Quick callback service should be a part of each subcontractor agreement. This is particularly important in plumbing and heating and air-conditioning agreements, which account for the largest portion of callbacks. An informal statistical analysis of homeowner complaints will help identify common ones for special attention. A good operating rule is: "The first complaint is the result of a mistake; the second is a coincidence; the third is a habit."

Bad habits should be rooted out of the construction process. Failing that, they can be recognized in a presettlement inspection checklist.

Design for Production

Principles

Reducing labor and material costs without reducing value or quality of the dwelling is the underlying principle of design for production. In other words, the dwelling is designed with the lowest cost per unit of worth to provide the highest value design.

The dwelling must be designed for both production and marketing. Design in effect, is a series of design compromises between building requirements and for selling requirements. Results must produce a product that will sell at a competitive price. Designing for production determines:
- How much the home will cost.
- Its appearance.
- Materials used.
- Equipment and facilities needed.
- How the home will perform.

Within reason, the principles of standardization, specialization, and simplification should prevail during the design phase. A reminder or check list, includes consideration of many items:

Market Facts (requirements and customer preferences)

Exterior Design

- Architectural style and exterior materials.
- Building shape and stories.
- Style and type of doors and windows.
- Foundation type (slab, basement, crawl space).
- Parking facilities.

Interior Design

- Floor plan, area, and arrangement.
- Number and size of rooms or spaces.
- Number and size of closets and storage.
- Interior surface materials.
- Design features serving as focal or functional elements (fireplaces, bars, cathedral ceilings).

Equipment and Appliances

- Heating and cooling.
- Kitchen and laundry appliances.
- Telephone, intercom and telephone connections.
- Special systems such as built-in vacuums, intercoms and security devices.

Dwelling Performance (in terms of adequacy and what is expected)

- Heat loss and gain (insulation, weatherstripping, window glazing, and storm doors and windows).
- Acoustical treatment and privacy.
- Structural durability.
- Operating costs.

Architectural Design

Modular Dimensioning and Standardization

- Use 2-foot and 4-foot major modular dimensioning with minor 4-, 8- or 12-inch dimension coordination to reduce waste and scrap and cutting labor.
- Use standardized building parts and pieces to provide maximum construction and design interchangeability without costly production changes.
- Use standard dwelling elements to achieve variation. For example, have standard 3-bedroom and 4-bedroom wings that can be used with several variations of the living area, or have a standard kitchen wall that can be used with such variations as placement of cabinets.
- Standardize, to the extent feasible, products by style, type, grade, size, color, pattern, or finish and eliminate variations where practical.

Simplify

- Develop maximum simplification of field construction drawings and details with sufficient dimensions, descriptions, or notes for foreman and workers.
- Plan for minimum lineal feet of exterior walls and interior partitions. Use uncomplicated layout and avoid unnecessary corners or offsets.
- Minimize or eliminate unnecessary parts and pieces and oversizing to reduce material waste and labor.
- Plan for simple gable-type roof system insofar as possible.
- Make a final examination of all plans, details, and specifications to avoid errors of omission and costly field construction correction.
- Set up a field-office feedback system to amend plans as needed to facilitate construction and clarify ambiguities.
- Evaluate house shape and configuration to determine whether simple designs might offer greater value.

Building Configuration and Design—Cost Considerations

Unit Plan Shape (1,200 sq. ft. average area)	Exterior Walls Lin. Ft.	Corners Out	Corners In	Roofing Areas	Roof System Gable Ends	Roof System Gable Valleys	Roof System Hip Hips	Roof System Hip Valleys
Square Plan	140	4	-	2	2	-	4	-
Rectangular Plan	142	4	-	2	2	-	4	-
Offset Rectangular Plan	156	6	2	4	3-4	-	NA	NA
"L" Plan	152	5	1	4	3	2	5	1
"U" Plan	172	6	2	6	4	4	6	2
"H" Plan	188	8	4	6	4	4	8	4

Building Configuration and Design—Cost Considerations shows the construction complexities of different home configurations. Each plan encloses 1,200 square feet of floor area. The square plan contains 140 lineal feet of exterior wall with only four outside corners and two roofing areas. The "H" plan requires 34 percent more exterior wall, eight more corners, four more roofing areas, two more gable ends, and four more roof valleys for the same amount of floor space. The probable optimum value shape in this example is the rectangular plan because floor and roof spans require less lumber than the square plan. Also, the rectangular plan may create a more "workable" room layout.

Engineering Facts

General Structural Items

- Design footing to balance bearing capacity of soil against live and dead loads. Determination of size based on soil capacity and loads will eliminate costly oversizing.
- Take vertical loads on foundation walls into account and/or reduction of backfill height to allow a reduction in foundation wall thickness.
- Check span reduction for beams or girders by adding extra column support for a possible total cost reduction.
- Check other stress grades, species, sizes, and spacings for joists and rafters to reduce material usage.
- Check all header loadings and size headers accordingly. Vertical stacking of identical-sized wall openings on load-bearing walls will reduce header size requirements. Headers can often be eliminated on nonload-bearing end walls. Eliminate all headers in nonload-bearing partitions.
- Use the no-cost design service usually offered by roof truss suppliers or truss plate manufacturers to achieve least costly truss configurations. For homes having 2- to 4-foot wall offsets, check cost of cantilevered truss designs using a continuous roof system.
- Specify and follow standard practice nailing schedules or product specifications to avoid over-nailing and reduce the chance of splitting.

Floor Framing

- Consider using a single-layer subfloor underlayment, glue-nailed to the joists for T-beam action.
- Select size and grade of joists on basis of maximum allowable clear span, and check 13.7-, 19.2-, and 24-inch joist spacing for possible material reductions.
- Plan for in-line (butt) joists to conform to modular dimensioning, stairways located parallel to joists, and mechanical systems located to avoid framing special openings.

Wall Framing

- When door and window rough opening widths are not to modular dimensioning, locate the opening so that one regularly occurring stud is adjacent to one side of the opening.
- Select smallest size and lowest lumber grade that will span header openings adequately. By raising the header to the top tie plate, header cripples and the second top plate can be eliminated.

Electrical

- Locate transformers as close to buildings as possible.
- Locate service entrance main panel as close to high current use centers as possible. Consider sub-panel for separated load centers.

Heating and Cooling

- Plan location of equipment, ductwork, and flues to avoid interrupting or displacing structural members.
- Use construction techniques, details, and products to reduce air infiltration.

- Minimize window area within market limitations.
- Use insulating glass or storm sash.
- Use insulated doors or storm doors.
- Increase thermal resistance of insulation.
- Shade or screen east and west glass areas.
- Reduce window height where feasible.
- Use light exterior colors.
- Increase natural attic ventilation.

Plumbing

- Check code variance possibilities for use of plastic materials for supply piping and DWV, reduced back-venting system, or single-stack reduced size vent system.
  - Check use of above-floor rear-discharge toilets and tubs to avoid displacing structural members.
  - Cluster and stack plumbing fixtures to minimize piping runs and number of fittings.
  - Consider shop-fabricated or site-fabricated plumbing wet wall.
  - Locate fixtures on one wall or use back-to-back arrangement in multiple bathroom.
  - Consider omitting bathroom window when using combination tub and shower.
  - Locate water heater as close as feasible to demand centers.

Case Studies

Foundation studies for the U.S. Department of Housing and Urban Development, National Forest Products Association, American Plywood Association, U.S. Steel, Alcoa, DuPont, and many others have documented that proper designing-for-production principles and techniques can reduce direct construction costs. Two case studies are included in this report to show examples of the significance of designing for production. These are the OVE System and the MOD 24 System.

The OVE System

The Optimum Value Engineered (OVE) building system, developed by the NAHB Research Foundation, Inc., for the Department of Housing and Urban Development, is an excellent example of design for production. Through the OVE System, the Foundation was successful in reducing direct construction costs by means of a systematic approach to dwelling unit construction by utilizing commonly available building materials and labor skills.

The OVE System is based on planning, engineering, and construction techniques applied to familiar building procedures. This means cost-effective planning and design principles are applied to overall design, structure, finishes, mechanical systems, and material and labor use. A prototype house, built to demonstrate the system, showed a savings of about 12 to 15 percent in direct labor and material costs when compared to conventional practices.

The system, as developed for the prototype house, was related to conventional wood frame and was based on a 2-foot planning module. For example, all framing members were spaced 24 inches on center. This means that joists, studs, and trusses were in alignment for structural continuity and simplicity of construction. Some members, such as studs and joists, were reduced in size through engineering design. Other members, such as foundation sill plates, were not used. Trim details were simplified throughout.

Four alternate cost-effective foundation subsystems were included to accommodate regional variations in climate, topography, and soil characteristics.

Cost-reducing optimum value engineering techniques applied to mechanical subsystems included a prefabricated plumbing wall, a simplified forced-air heating system, and a preassembled electrical wiring harness.

Of the total direct cost savings for this small home, 60 percent was for material and 31 percent for labor—a fairly typical material-labor ratio in home building at the time the system was developed. Framing and related items, such as sheathing and siding, were the largest major categories of savings and represented 65 percent of total savings. This was followed by foundation savings of 13 percent, mechanical savings of 11 percent and other savings of 11 percent. The "other" category included such items as interior paint and interior trim.

Based on the builder's records, the savings in direct construction costs represented more than 12 percent of typical direct construction costs for a conventional home of this type. The reduction in labor and materials costs would allow these resources to be used to build approximately one additional home for every eight presently built.

The Engineered 24 Framing System

The Engineered 24 System home (originally called the MOD 24) consolidated a series of important construction advances made over a period of years. Less lumber was used in the framing operation for floors, walls, and roof by spacing framing members 24 inches on center than for typical designs.

In both the Engineered 24 System and the OVE System, such basic framing members as trusses, studs, and joists were aligned to form a series of in-line frames so that all concentrated loadings were in alignment. NAHB Research Foundation industrial engineering field studies revealed a cost savings in the range of $200 a house (labor and material) with this system of building.

The Cost Buster Houses

The two Cost Buster houses demonstrated that construction costs can be reduced without compromising health and safety. The City of Las Vegas made special exceptions to the building code to allow this demonstration of modern, economical construction to be free from unnecessary codes and regulations. The Cost Buster concept was developed by the builder and the NAHB Standing Committee on Research. It drew on past research done for the U.S. Department of Housing and Urban Development by the NAHB Research Foundation. By utilizing a large number of cost-saving features, this project achieved a 25 percent reduction in direct costs compared with a typical Las Vegas house of similar size.

The Approach '80 Project

The Approach '80 Project extended the Cost Buster concepts into the land development field. It was sponsored by NAHB's Land Development and Research Committees with HUD research funding. The objective was to produce a safe, healthy environment inside and outside the home and to provide housing that would be a positive addition to the neighborhood, socially, economically and aesthetically. Seven of the 38 Approach '80 units were demonstration houses. Their construction systems and components provided alternatives to conventional construction. Alternative techniques included Optimum Value Engineered framing, reduced concrete slabs and footings, all-weather wood foundations, underfloor plenum heating and cooling, plastic plumbing supply and DWV systems, and light gauge steel interior partitions. The Approach '80 units included 10 detached houses, 7 duplexes, 2 threeplexes, and 2 fourplexes ranging from 704 to 1,116 square feet. Average savings per unit was $5,491 or 10.6 percent compared with conventionally built Las Vegas homes.

Material Cost Control and Reduction

Principles

Many principles and techniques for material cost control and reduction are identified in the section on "Design for Production." The purpose of this section is to identify general principles and methods of material cost control and reduction. Some specific illustrations of these are also included.

Evaluating material cost control potential requires an examination of the interrelation of all costs. Material, manpower, machine, management, and money costs are all covariables. A change in any one may mean a change in any or all of the others. The first and possibly most productive places to look for material cost control are in the design and purchasing phases.

Purchasing

The cost of material is usually the single largest dollar amount in the sales price of a home, making it the most important area for reducing and controlling costs. On the average, the selection and purchasing of materials can be improved to effect as much as a 10 percent savings on material cost. A 10 percent savings on materials may represent as much as 4 percent savings on each sales dollar.

Where is this potential improvement? A part of it, of course, is in direct savings brought about by substituting materials or seeking better market prices. Other substantial savings are found in improving purchasing practices to reduce inventory costs, storage costs, delivery delays, pilferage, and waste.

Five basic steps are involved in the purchasing function, all of which need to be built into an effective purchasing system. These are Product Identification and Evaluation, Source Selection, Ordering, Followup, and Receiving.

Product Identification

Product identification describes what is to be ordered. Production and purchasing departments routinely make these determinations. However, engineering and sales departments are also involved, particularly in evaluating new products.

Material evaluation should be a continuous function. Changing conditions of technology, availability, customer preferences, and price structure can make last year's decision a costly one this year. Materials should be reviewed periodically with special emphasis on high-ticket items and those with unstable market conditions. Periodic evaluations break up the inertia of potentially wasteful habit buying.

New products must also be evaluated in competition with present materials. A course must be steered between rashly trying new products and lagging several years behind competitors in adopting new products. Suppliers, salesmen, and trade magazines are primary sources of new product ideas. Some larger-volume builders establish formal new-product committees with representatives from the engineering, production, sales, and purchasing departments.

Product evaluation falls into two groups of considerations: cost and non-cost. Cost considerations do not start and end with invoice price. The all-important price to be evaluated is in-place cost. This includes invoice price, installation cost, replacement, repairs, and even consideration of follow-on warranty responsibility. Further, today, operating costs are an increasing customer concern, especially in regard to energy use.

In a recent Foundation study involving self-drilling drywall screws, one bargain lot included 6 percent deformed screws. The lost rhythm, gun jamming, damaged drywall sheets, and screw-removal made them a costly buy at any price, and the cost consideration was not, of course, reflected in the supplier's price.

Non-cost considerations include utility, quality, appearance, customer acceptance, and maintainability as well as supplier services. They effect the real value, and sometimes the market value of the house, but cannot be charged with a specific in-place cost differential.

Quality also may vary over time. A manufacturer may allow his product to degrade and only periodic reevaluation by the purchaser can guard against this.

Premium-cost items should only be specified if they contribute a commensurate amount to the value of the home. Premium material cost should be offset by either a saving of in-place cost, an increase in market value of the house, or in some instances adequate reduction in operating or maintenance costs. Unnecessary premium quality adds only higher cost.

Frequently, an important value consideration is the purchase of brand name items. Well-advertised or reputable brands may be worthy of a premium price if customer acceptance, advertising tie-ins, or better performance improves sales.

Source Selection

Once the product and alternatives have been identified by specifications and quantities, a supply source must be selected. Source selection is principally a cost comparison. It is, of course, much more than a simple weighing of price differentials. Cost comparisons are complicated by the fact that the prices include material plus supplier services, not just material cost alone.

Full cost includes all services required to install the material in place and even extends to replacement or repair under warranty. These costs are distributed among the material manufacturer, distributor, supplier, subcontractor, and builder. Costs for transporting, financing, storing, protecting, and stocking inventory are paid in the long run by the final purchaser of the material. Whether these costs are hidden in the supplier's price or performed by the builder's own workers, they should be recognized as legitimate and usually unavoidable.

Distributor's or dealer's services have acknowledged values that are considered in price evaluations. Services rendered by dealers may include:
- Providing market information
- Advising on new or improved products
- Placement at site to minimize site labor
- Planning and estimating
- Carrying inventory
- Precutting
- Bundling, palletizing, or strapping
- Subassembling
- Delivering on schedule
- Financing

In return, the builder pays a price that includes an unspecified markup to cover the services rendered.

The supplier's pricing structure can be an important consideration in choosing a source. For example, taking a purchase discount of "2 percent 10 days, net 30 days" returns a 2 percent saving by paying 20 days earlier. This is a 36 percent annual return, assuming payment in 10 days. In a tight money market, suppliers often discontinue such discounts.

Compare the supplier's discount and penalty rates with the cost of construction financing. Discount taking may be of limited saving value when the builder's own interest costs approach the discount rate. At this point, maintaining a good credit rating may become the principal incentive for making timely payments. Credit terms are frequently less a matter of supplier policy than of negotiation between builder and supplier. By recognizing mutual problems and negotiating well, the builder may realize a substantial saving.

Subcontractor-installed material may be either builder-furnished or subcontractor-furnished. The determination is usually based upon relative bargaining position of the two parties with suppliers. The builder may reduce lien exposure by furnishing materials. Psychological aspects of this decision may be equal in importance to the economic factors. If a subcontractor furnishes his material, he is likely to take a more personal interest in controlling waste. On the other hand, the builder can lose control if he assumes that "material is the subcontractor's worry."

The "heart" should be where the money is and that includes subcontractor-furnished as well as builder-furnished material.

A qualified supplier should maintain sufficient stock on hand to deliver on schedule and to fill emergency material shortages. If uncertain, a visit to potential supplier's warehouses can quickly determine adequacy of their inventory stocks and distribution systems. It is also an opportunity to meet the warehouse supervisor on a first-name basis.

Purchasing "big ticket" materials direct from the manufacturer can be advantageous. With sufficient volume of production, project quantities, annual supplies, or carload lots can be purchased. Volume purchases may result in dollar savings and sometimes help establish priority preference for items in short supply. However, financing, warehousing, handling, and other costs may offset the apparent savings. Materials often purchased direct include lumber, plywood, sheathing, roofing, windows, doors, appliances, drywall, and carpeting. Some builders have established their own building material dealership so as to have better materials and cost control and possibly to effect a small cost reduction.

In addition to quantity discounts, several other factors operate to establish material prices. Market fluctuations, short supplies, delivery costs, quality levels, servicing, and financing arrangements are important. But the most significant factor is negotiation—according to many builders, an important pricing consideration in nearly all building material markets.

Even though volume purchasers can realize price advantages, an actual saving may not always be made. Many builders buying in carload quantities believe that their indirect cost increases related to volume purchases about offset the price advantage of volume purchases.

Inventory turnover rates are interrelated with volume purchasing decisions. Larger order quantities mean larger inventories and more funds tied up in stock. Material represents money, and material in storage represents dormant money. Average inventory costs become a permanent capital investment. Only a faster turnover rate can reduce these "impounded" assets that reduce operating capital.

A number of factors enter into the determination of optimum rate of turnover. Perhaps the most troublesome is the popular, but fallacious, assumption that a single rate of turnover applies to an entire inventory. In reality, a separate rate of turnover applies to each purchase in the inventory. The rate of turnover is actually an average of many individual rates. Rate of turnover for a commodity is based on annual usage divided by average inventory. Turnover rate as an average of individual commodity rates provides the clue to an effective control system used in manufacturing for many years. This system, popularly called the ABC Method, was based upon work done by the General Electric Company. The ABC Method is as adaptable to home building as to other manufacturing since building is essentially a manufacturing process and the product is a house.

The ABC Method is based upon the singular phenomenon defined by Pareto's Law. A builder's inventory is composed of a mix of costly and inexpensive materials. Some 20 percent of the commodities will represent some 80 percent of the dollar value; the remaining 80 percent represent only 20 percent reduced potential. Many organizations expend similar effort in purchasing and storing all items, giving "equal treatment" to all. The ABC Method uses three categories: "A," "B," and "C." The first task in the ABC Method is to tabulate inventory items in the order of their yearly dollar cost and then to select a few of the top-dollar value commodities as "A" items. The number of "A" items will be gradually increased up to perhaps 10 percent as controls are effected on the top items. The next step is to select the relatively cost-insignificant "C" items from the bottom of the list; these will probably constitute 50 to 70 percent of the total number of parts. The remaining items are average-cost "B" items.

"A" items should then be subjected to thorough cost analysis including:
- Study of minimum technical requirements for adequate performance.
- Competitive materials and attention to new product development.
- Alternate sources of supply and familiarity with market conditions.
- Volume/price breakpoints.
- Possibility of high turnover rate to keep down inventory investment.
- Tight scheduling to reduce storage time.
- Close expediting of orders.
- Accurate record keeping and inventory controls.
- Competitive pricing practices such as annual contracts, direct purchases, price negotiations, and sealed bids.
- High management interest.

"B" items follow a normal purchasing routine.

"C" items represent relatively little dollar value so that their controls can be quite loose, including:
- Infrequent purchases (unless storage is bulky).
- Reserve stock for standard items, reordering when reserve is opened.
- Bulk or breakpoint quantity purchases for standard items.
- Minimum purchasing department effort.

The ABC Method offers a practical approach to singling out items deserving special emphasis in purchasing and controlling. It indicates, also, those items on which effort expended would generate maximum and minimum return.

Factors entering into source selection decisions are not always logical. The most prevalent nonlogical factor is habit buying. The way we buy materials can be as habit forming as the way we button our coats. Becoming a captive customer through sheer inertia is easy. On the other hand, breaking the habit by periodically comparing other sources is relatively easy.

One insidious form of source selection is the kickback. Bribery is not represented only by political scandals or the biblical faithless steward. This practice is prevalent in industries where large contracts are handled. The kickback may range from a bottle of whiskey to a percentage of the contract. In any event, it represents money taken from the company's pocket. It can be detected only through comparative shopping by persons outside the normal purchasing channels. The three-bid system and trying new suppliers is a help but not a cure.

In times of inflation, price forecasting becomes a very chancy business. The market shifts from fixed prices and assured supplies to escalating prices and unpredictable supplies. The old cost-estimating method of adding up prices changes over to a process more akin to handicapping a horse race. Accent is no longer on estimating future price fluctuations but on attempting to estimate the future rate of price increases. Some builders try to protect against short-run lumber cost increases by going into the lumber futures market as a hedge. Others use annual usage, fixed price, noncancellable contracts to entice manufacturers or distributors to guarantee short-run prices.

Ordering

Material should be ordered on formal written purchase orders. Reducing requirements to writing decreases the chance of error in both buyer's and supplier's organizations. In addition, purchase orders are necessary for legal protection in the event of later disputes. Purchase order forms should include standard terms and conditions of the contract. A written purchase order provides a memorandum for the builder's records indicating that the material has been ordered, provides protection against fraudulent orders if specific signatures are required, furnishes an accounting record of the contingent accounts-payable liability, and further provides a list for checking out receipt of the material.

The extent of standardized material requirements of home building allows use of standard forms for purchase orders. Material descriptions, specifications and delivery instructions can be filed on cards to be reproduced by photocopy directly onto the purchase order, or they can be filed in computer memory to be computer-typed onto the purchase order.

In the case of standard house models, a bill of materials can be filed with the supplier so that individual house kits are ordered by reference, or the bills or materials can be photocopies for each separate kit. In either case, the repetitive work of preparing detailed purchase orders is avoided.

The standard order is checked for accuracy on the first house. Subsequent orders are then written against a proven bill of materials. This method of ordering, unfortunately, favors development of habit buying. To break up the habit pattern, a program of periodic material reevaluation should accompany the use of standard orders.

Followup

Expediting purchase orders is a form of crutch. The ideal way to do it is *don't*. Suppliers will learn to rely on the purchaser's expediting efforts instead of their own. If the supplier has a competent organization, it can be trained to do its own followup. If the organization is not competent, it is not providing a necessary service. These are some hints on training the supplier to meet delivery schedules:

- Include a tight delivery schedule clause in your purchase order form.
- Assure that your supplier receives at least the minimum leadtimes that he needs for filling your orders; reaffirm schedules 24 hours before delivery if desired.
- Request advance telephone notification of material back-ordered or unavailable.
- Make clear to the supplier that he must meet your delivery schedules within mutually agreed-upon tolerances.
- Reproach the supplier immediately, fairly, but firmly on any missed schedule. The object is to convince your supplier's staff that your delivery schedules are sacred and that a missed schedule will bring an immediate complaint.
- Make sure that the receiving site is ready to receive the delivery on schedule.

Receiving

When materials are delivered, they are inspected for quantity, quality, and condition. Discrepancies are marked on the shipping copy of the invoice, which is compared to the purchase order before being authorized for payment. Discrepancies or shortages are followed up by the purchasing department.

The supplier's delivery responsibility does not end with getting materials on site. Delivery should include protective stacking in specifically designated locations. These advantages come from proper delivery and storage of materials:

- Protection from the elements by stacking on skids, covering with protective sheets, storing under roof.
- Protection from pilferage by steel banding, storing indoors, or storing in relatively inaccessible locations (for example, storing pilferage-prone shingles on the roof).
- Reduced handling cost by storing materials near point of use, such as foundation block inside the excavation at several locations, gravel piled inside the excavation, or gypsum wallboard distributed inside the house.

If the supplier is to provide careful stacking of delivered material, the buyer has related responsibilities. He must provide a storage site that is cleared, well drained, supplied with skids to keep material off the ground, and prominently marked. He must notify the supplier of desired date well in advance. The simple expedient of posting prominent directional signs and site identification numbers can give a businesslike appearance and put delivery drivers in a cooperative attitude.

Material Handling

Material use always starts with handling. Proper handling techniques can reduce breakage and pilferage as well as reduce associated labor costs. A good material handling program consists of:

Delivering the Right Amounts

- Having lumber precut to exact length when feasible.
- Delivering *only* the amount needed so that any shortage reported will serve as a monitoring device for a problem needing investigation.

Delivering to the Right Place

- Good material handling means material is *not* handled.
- Place it where those who use it will handle it least.

Delivering at the Right Time

- Deliver according to sequence of use so that placement does not interfere with access to other material or the work of other trades.
- Be sure materials are there when needed because manpower waiting costs quickly become excessive.

Good material handling also includes good housekeeping. A clutter of materials at the work place tends to cause waste, pilferage, and lost time searching for materials. Wherever possible, shipments to the job site should be strapped or otherwise unified.

A clear method of fixing accountability for material control is important. This means giving one man both the responsibility and the opportunity to monitor material disposition. The site supervisor should be in charge of materials. He may not be responsible for ordering materials, but he should be responsible for seeing that the correct quantity and quality are delivered and that the materials are used as intended. If more materials are needed, he should have a method for requisitioning replacement materials. If materials are left over, they should be transferred to another site or warehouse and credit given to the house where the excess occurred.

Sample Requisition, Transfer, and Credit Form can be used to control requisitions, transfers and credit.

Scrap and Waste

Scrap may be defined as unusable materials such as short end-trims, small cut-outs or, in general, materials that have no salvage value when labor costs for salvage exceed material savings. Waste may be defined as discarded materials that have salvage value or misused material or materials damaged beyond use by carelessness. Materials considered scrap by some might actually be waste because, if properly controlled, the scrap would not have been generated in the first place. Also, the belief that it costs more in labor to salvage materials than the value of the materials often creates false scrap. For example, workmen often cut blocking from full length pieces of lumber rather than from shorter pieces because it takes less time. The unused short pieces become "false scrap" if they could have been salvaged profitably.

Reducing scrap and waste means making control a standard operation, using such steps as:
- Placing racks or bins for different sizes and grades of materials onsite. (The racks should be designed for easy portability by forklift or otherwise.)
- Monitoring scrap and waste disposal. (Have a system of identifying the amount and disposition of scrap and waste. Hold one person accountable for scrap and waste disposition.)

Sample Requisition, Transfer, and Credit Form

☐ Requisition
☐ Material Transfer
☐ Credit Memo

Date _____ No. _____

For Office Use Only

To:
  Lot No. _____
  Subforeman _____
  Phase Foreman _____

From
  Lot No. _____
  Subforeman _____
  Phase Foreman _____

Date	Description	Quan.	Unit	Unit Price	Total Price

TOTAL

Signed _____
Received by _____

In one specific instance, an industrial engineering study by the Foundation determined that a major East Coast builder was increasing costs substantially by not controlling scrap and waste and by not paying close attention to material use. The study revealed the framing crews were:
- Framing openings that did not fit revised window sizes.
- Prefabricating and discarding as many as four extra headers per house.
- Using large structural headers in nonload-bearing walls.
- Using studs under breaks in top plates instead of having break fall on regular stud spacing.
- Using bridging that was not required or necessary.

Also, an inventory of scrap, waste, and damaged materials showed that trusses were broken, extra headers were lying about, a load of studs was used as a truck ramp, over 100 sheets of plywood and sheathing were damaged beyond usefulness and many window frames were broken.

On 53 lots, total wastage amounted to about $8,000 in materials alone, and this did not take into account labor costs always tied to material waste and misuse. Nor did it account for materials taken home by workmen, one of whom was understood to have been building an addition to his house from "scrap" materials taken from the jobsite.

Such waste can be almost entirely avoided by better control procedures. Since materials are in specific sizes and quantities, they can be counted and measured. By maintaining strong material control, some control is also gained over manpower, because manpower costs usually bear a direct and fairly constant relationship to material costs.

Material Use

Good material use should begin at the drawing board. Excessive materials are often used because of tradition, misunderstanding of structural adequacy, a certain concept of quality or, in many cases, because the code requires it. Certain principles regarding use of materials are relatively easy to put into effect without disrupting present methods. Other principles must be weighed to determine whether tradition, quality, or the building inspector's attitude can be modified.

Modular Framing

In modular framing, the design and spacing of framing members fits the manufactured sizes of covering, insulation or other materials that must fit the framing.* This minimizes cutting labor and reduces scrap and waste to a minimum.

Most cutoffs should be modular with the framing—that is, of a size which is a whole number multiple of the framing spacing—and can likely be used elsewhere on the house or another house. Make framing modular out-to-out of framing. The Modular Design sketch shows an ideal that is seldom—if ever—reached. This model produces no scrap except for door and window openings and possibly gable ends. Minimal cutting is made.

To preserve the module, the following steps must be taken:
- Maintain specified dimensions.
- Order tongue-and-groove plywood specifying a face width of 48" or else it may be cut from a regular sheet, making the face width 47⅝".
- Make whatever minor adjustments are needed in spacing for framing to allow the recommended gap between sheets (anywhere from 1/32" to 1/4") depending upon location and moisture conditions.
- Use in-line joints for subfloor to save cutting.

---

*See *Unicom Manual* 1 and 2, National Forest Products Association, Washington, D.C.

In addition to these specific cost-control techniques, these general considerations should be kept in mind:

• When ordering and storing lumber, determine the right combination of spacing, species, size, and grade for cost-effectiveness. Besides 12, 16, and 24 inches o.c. spacing, 13.7 and 19.2 inches are modular to 8-foot sheathing. Sometimes using different spacing will allow use of a lower grade or smaller size joist or rafter. Using a larger size joist or rafter may allow wider spacing or use of a more economical species or grade. Alternative combinations should be evaluated for overall cost-effectiveness.

• Inspect material upon receipt, then protect it until used. Not only should size, quality, and quantity be checked against the order, but such things as moisture content should be determined. If moisture is high, lumber will shrink and may warp. Check the ends of framing materials for "heartwood" pieces. If many hearts are noticed, it may indicate cuts were made from smaller trees. This can mean warping after installation which can be expensive in material and labor to correct. Lumber should remain bundled and protected from moisture until used.

Additional cost advantages can be gained by considering alternative materials and techniques.

• Use a cost-effective center girder. This means engineering the design of the girder no matter what material is used.

• Engineer the entire structural system. Modify the floor framing to fit the design load. For example, sleeping areas may require only a 30 lb/sq. ft. live load versus 40 lb/sq. ft. in other parts of the house.

• Glue-nailing of subfloor to joists may allow increased joist span, smaller joists, or increased spacing.

• Know which walls are load-bearing and which are nonload-bearing, and use only the materials you need for each. Eliminate structural headers in nonload-bearing walls and in some basement openings. The floor band joist may suffice as a basement opening header.

• Consider single-skin coverings on exterior walls and floors. Single-layer sidings without sheathing and combination subfloor/underlayment materials reduce cost and are used successfully by many builders.

• Look at the whole assembly in determining thickness, species, grades, etc. For example, when wood strip flooring is applied perpendicular to floor joists, the subfloor may be reduced in thickness or possibly eliminated altogether.

• Eliminate soffits, excessive trim, etc., where feasible.

Modular Design House

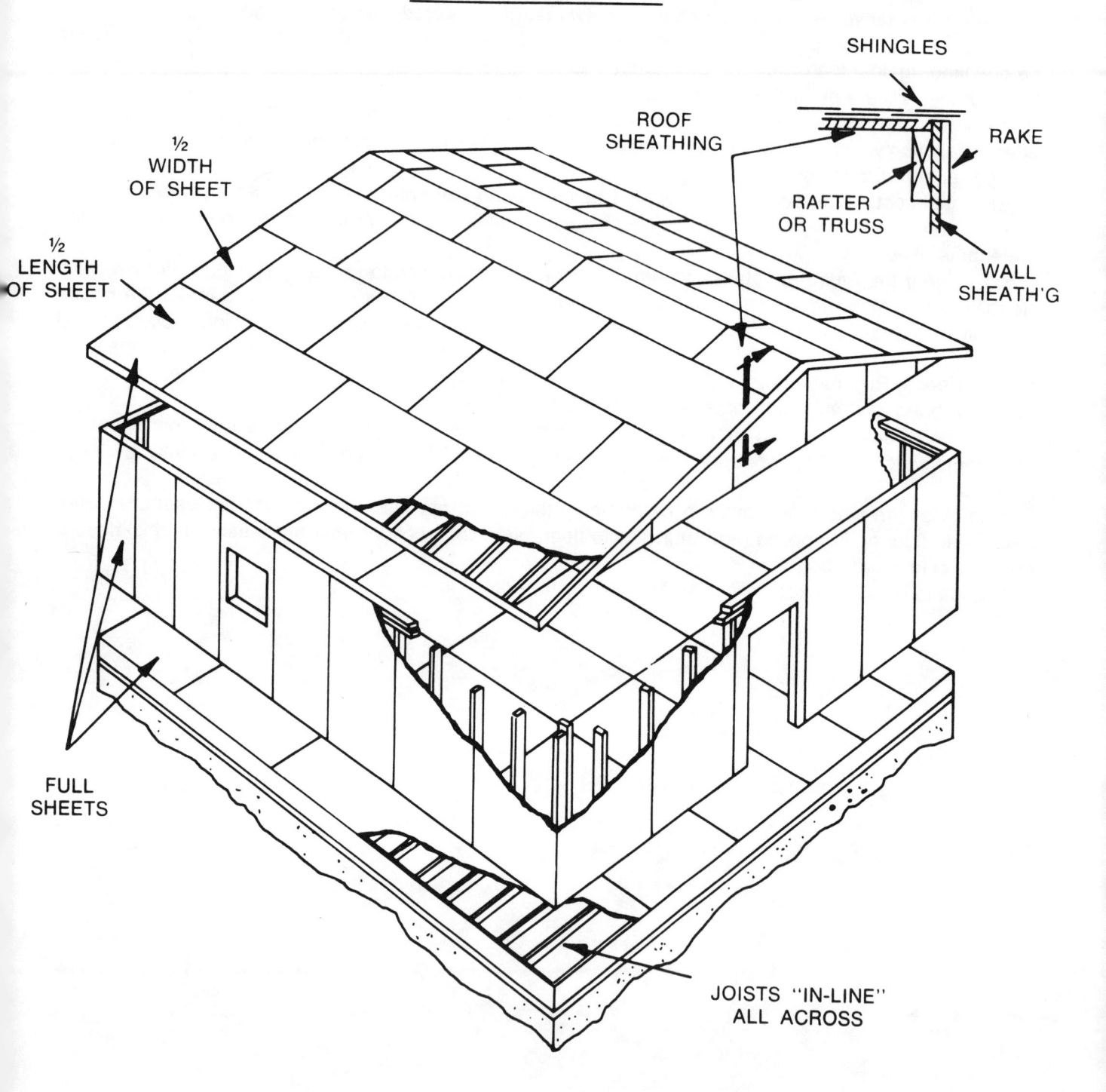

In-Line Load Support

Use "in-line" load support where feasible, as, for example, in overall framing.

STANDARD FRAMING

Truss, stud, and joist may or may not be in line.

MATERIAL-SAVINGS FRAMING

Truss, stud, and joist are directly in line.

Load is supported more efficiently.

In addition to savings on top and bottom plate and band joist, structural headers may be omitted.

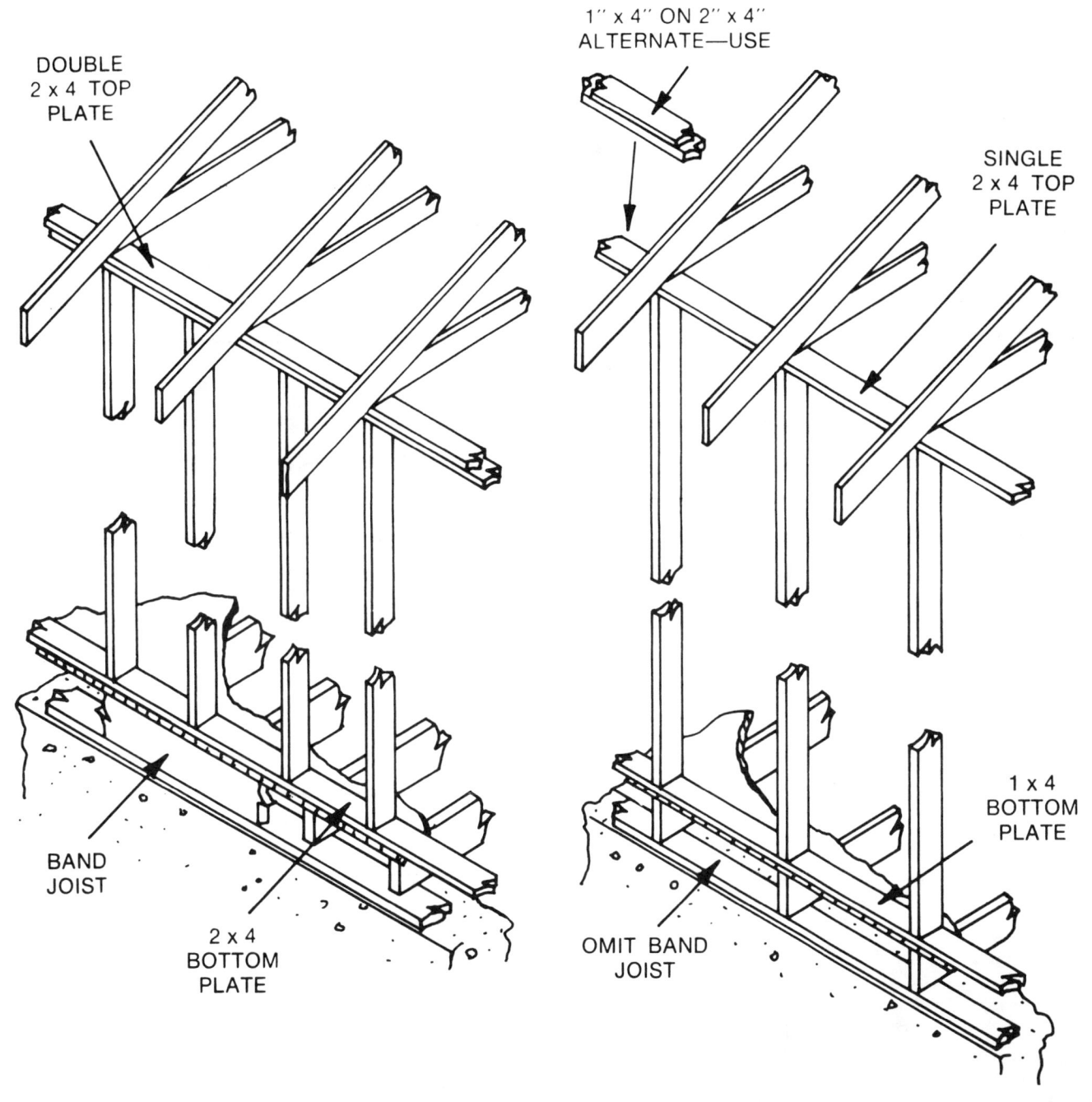

DOUBLE 2 x 4 TOP PLATE

1" x 4" ON 2" x 4" ALTERNATE—USE

SINGLE 2 x 4 TOP PLATE

BAND JOIST

2 x 4 BOTTOM PLATE

OMIT BAND JOIST

1 x 4 BOTTOM PLATE

45

Savings by Use of In-Line Joists

STANDARD

MATERIAL SAVINGS

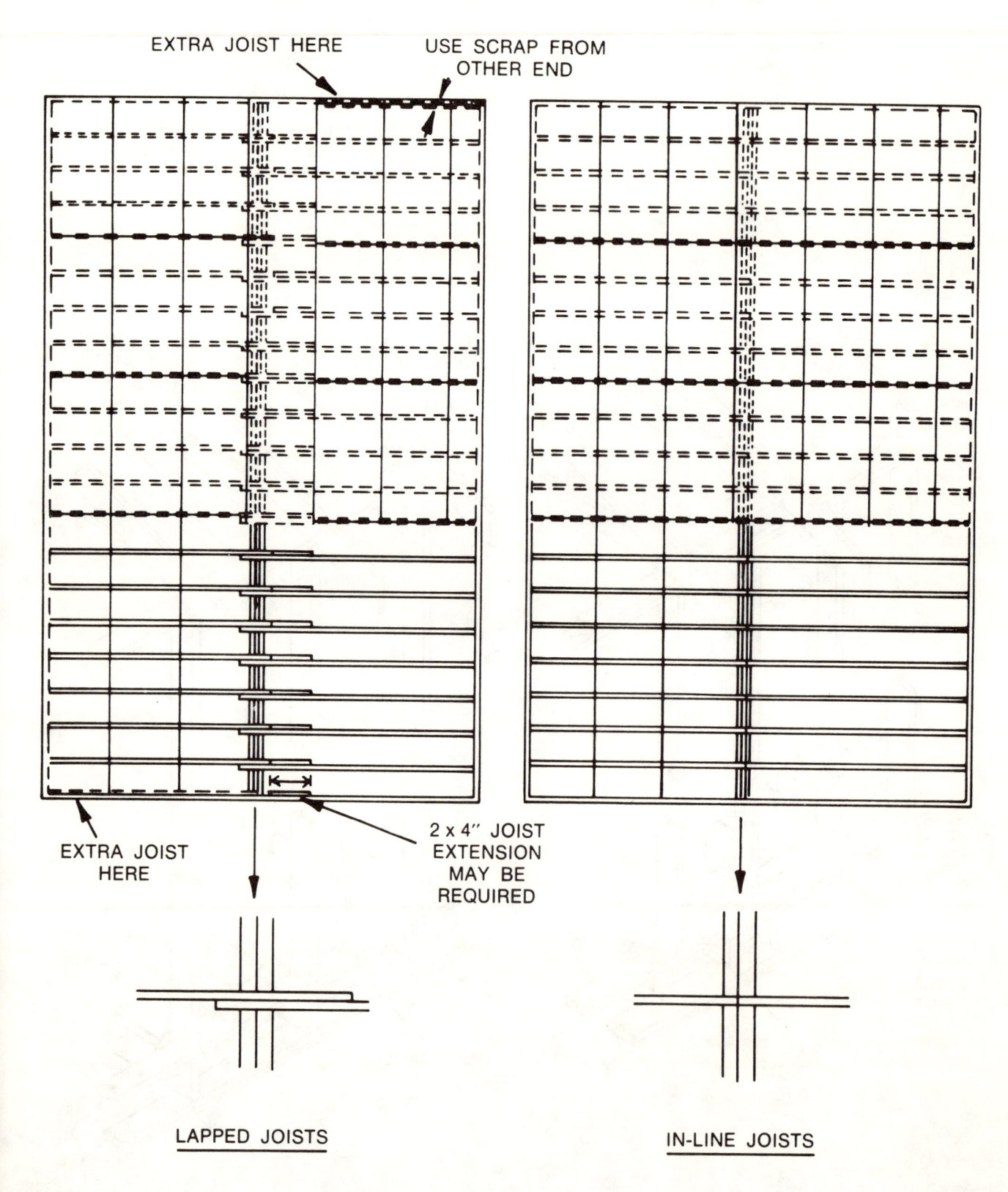

EXTRA JOIST HERE

USE SCRAP FROM OTHER END

EXTRA JOIST HERE

2 x 4" JOIST EXTENSION MAY BE REQUIRED

LAPPED JOISTS

IN-LINE JOISTS

Stairway Fitted to Joist Layout

Less joist material will be required if the stairway opening fits the joist layout. Also, cuts of subflooring will be modular with the framing, making the scrap more likely reusable.

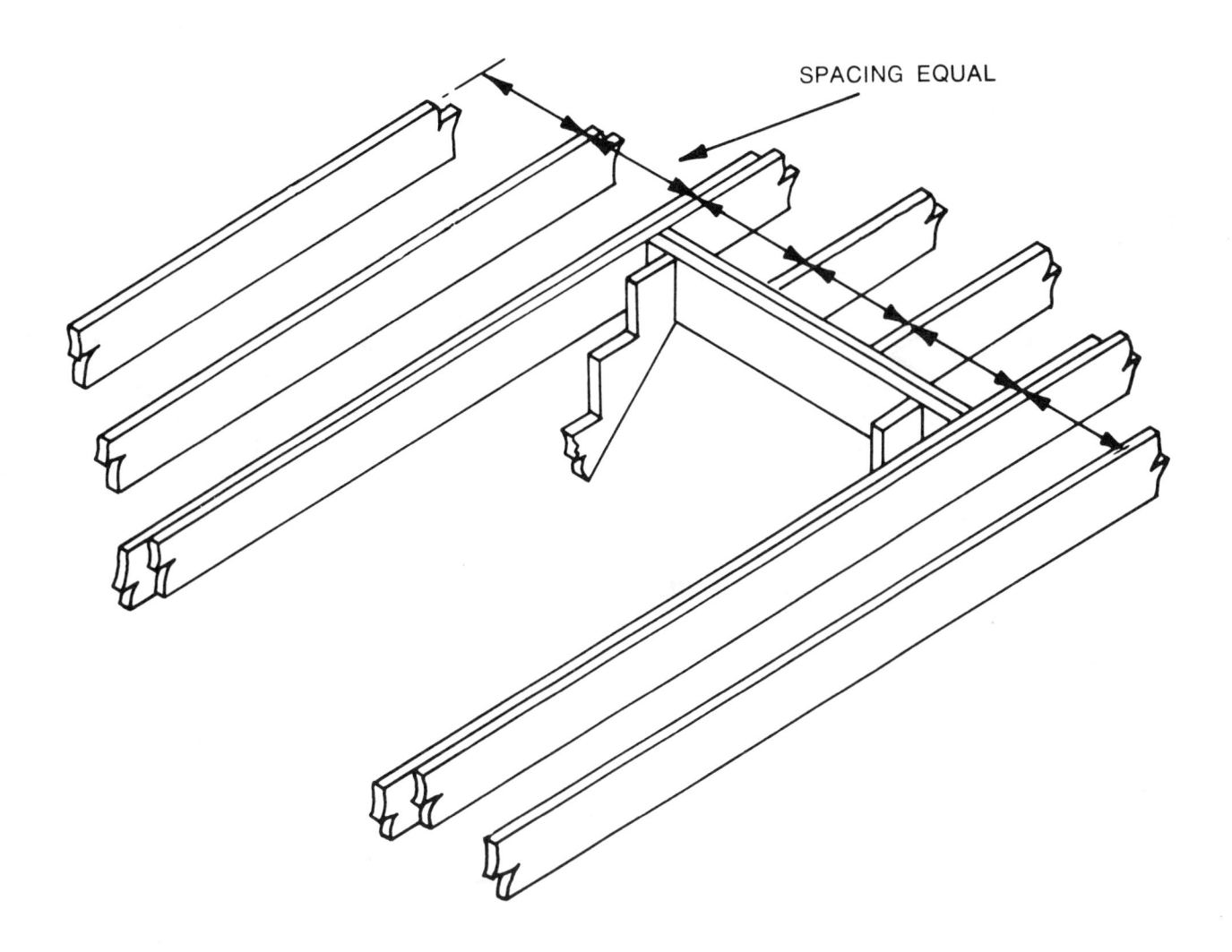

Layout and Cutting Schedule for Gable Roofs

Determine the best layout for covering materials before cutting. From this layout make a cutting schedule. (A Foundation study on roof sheathing costs indicated that about ⅔ of the waste from unplanned cutting can be saved by judicious preplanning.)

Partitions Over Joists

All else being equal, savings can be effected by placing partitions over regular floor joists; avoiding an extra joist. Double joist should be used only when required by local code.

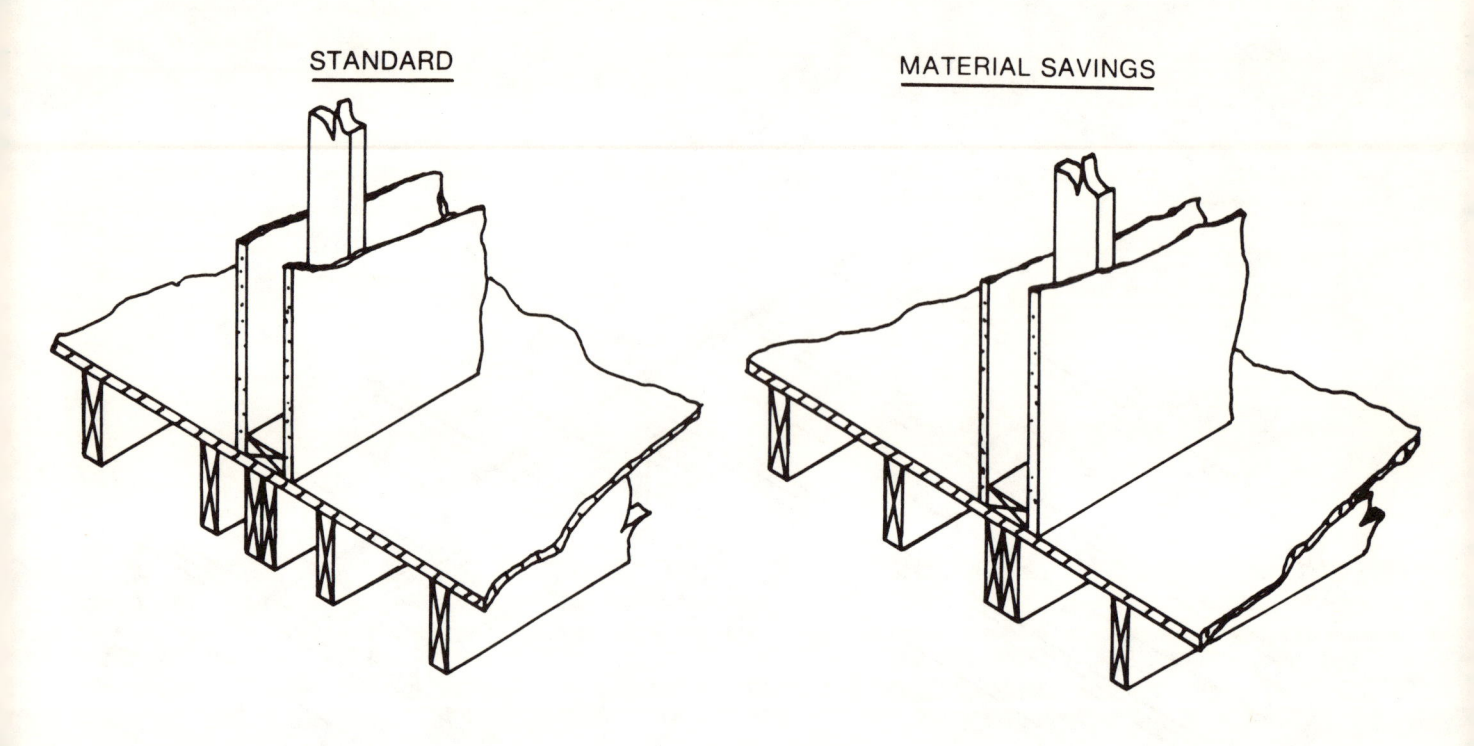

STANDARD

MATERIAL SAVINGS

Openings Fitted to Existing Framing

Material savings can be made by fitting openings to existing framing, as, for example, in a window opening.

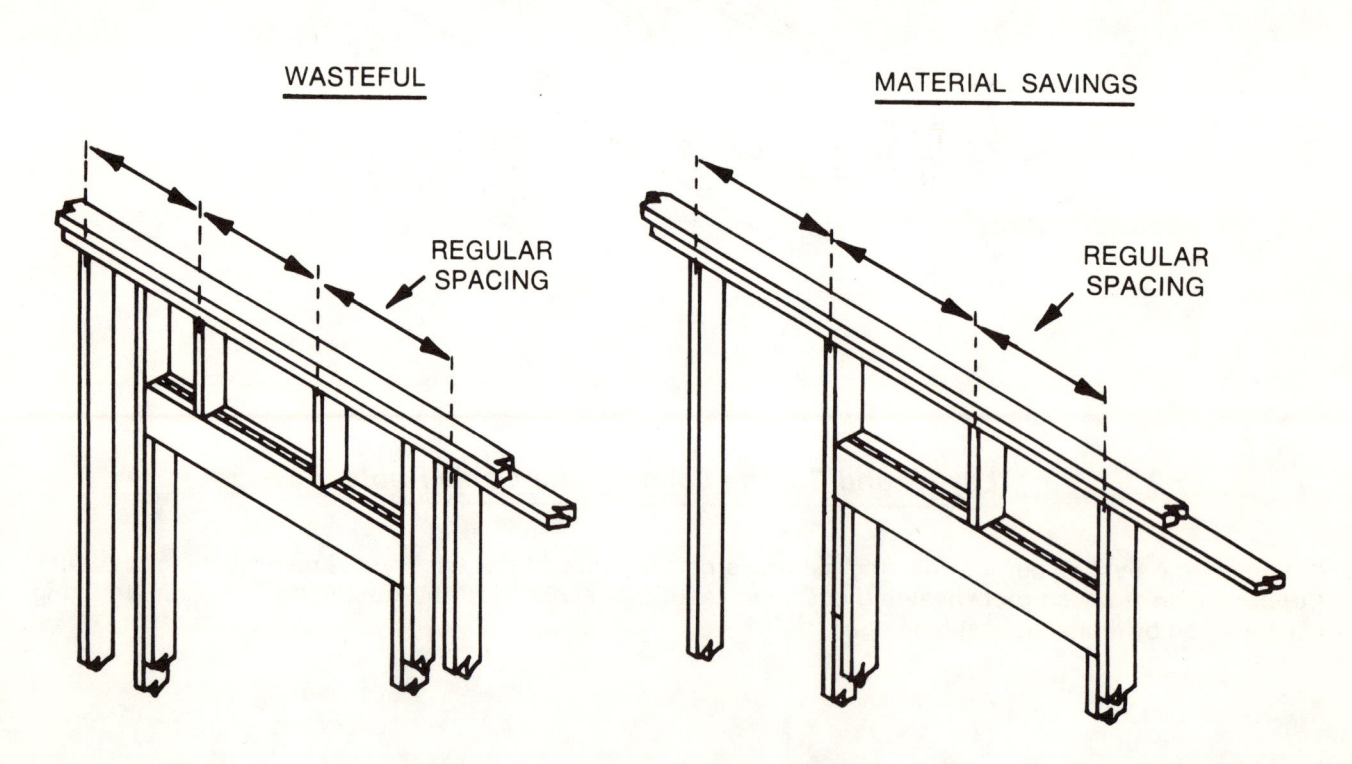

WASTEFUL

REGULAR SPACING

MATERIAL SAVINGS

REGULAR SPACING

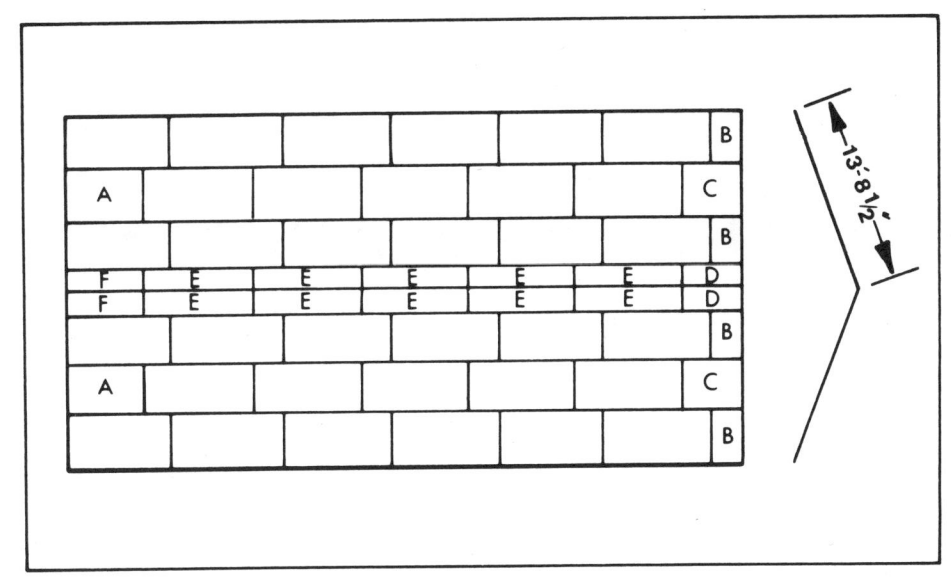

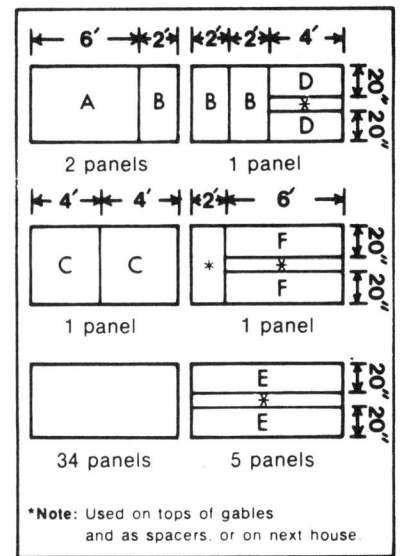

Typical Gable Roof Layout—4 in 12 slope 50' x 27'-5" trusses 2'-0" o.c.

Cutting schedule for gable roof at left

*Note: Used on tops of gables and as spacers, or on next house.

Layout and Cutting Schedule for Hip and Valley Roofs

Key dimensions determine spacing of first jack in a hip roof, first rafter or truss in roof with valleys. Modify "key" dimension for efficient material utilization.

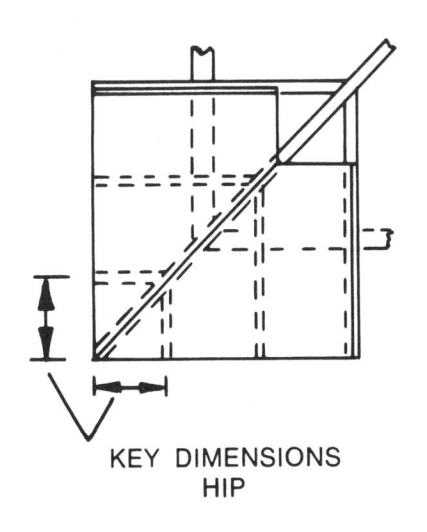

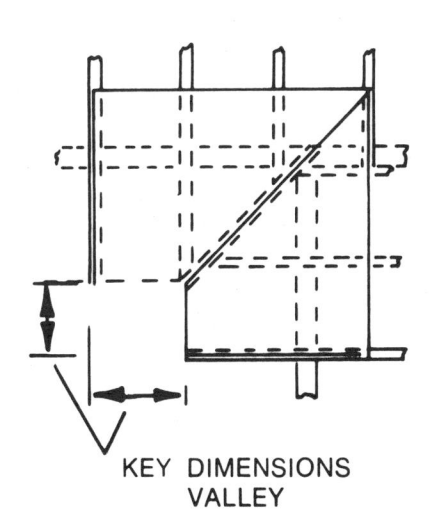

KEY DIMENSIONS HIP

KEY DIMENSIONS VALLEY

Labor Cost Control and Reduction

Labor in the residential building trades is more efficient and rationally organized than popularly thought; nonetheless, ways can usually be found to improve labor productivity.

To expect that the key to improved labor productivity can be found in increasing the worker's effort is natural, but onsite industrial engineering studies show that an acceptable amount of effort is usually applied to tasks at the construction site. The individual labor effort, however, is often hampered by poor methods, inadequate tools and equipment, improper distribution of skills, or onsite delays.

The principles of work improvement do not necessarily aim to make people work harder. In fact, a good work improvement program will probably mean that less physical effort is expended because the men "working smarter, not harder."

The principles of work improvement analysis actually have their historical roots in the construction industry. Frederick Taylor, Frank Gilbreth, and D.J. Hauer, the fathers of modern industrial engineering techniques, developed and tested their work improvement methods on the construction industry. Yet, paradoxically, almost every other industry except construction has accepted and used these techniques to improve productivity. Since construction managers consider their work nonrepetitive and their product one that differs from project to project, they perhaps believe that improvement methods used in factories do not apply.

Although the end products may be different, individual tasks that make up the products are indeed repetitive and are prime candidates for work improvement methods. Also, the subcontractor system of awarding contracts after competitive bidding is believed to be a method of assuring efficiency. With the subcontract system, the contract may be let to the least inefficient of the subs bidding. However, the best possible construction at the lowest possible cost is not a guaranteed result of competitive bidding. Every dollar lost by the subcontractor because of inefficiencies is usually lost by the builder in the long run. Every penny saved by improved methods increases the subcontractor's profit and eventually benefits the builder through lower bids or more reliable and stable subcontractors.

The responsibility for low productivity falls not only on the worker involved but, more particularly, on his direct supervisor. Most of the loss of efficiency is due to involuntary idleness or poor methods causing delays for the worker. The major causes of delays are concerned with delivery of materials, scheduling of work, incomplete instructions, unbalanced crews or crew size too large, rework, and general lack of specific planning.

The site supervisor plays a major role in controlling direct onsite costs. Although final responsibility for success or failure of a control program rests with the builder, the site supervisor is the intermediary who makes the control program work.

Labor accounts for about one-third to one-half of the total direct construction cost, and material accounts for the rest. Onsite labor and material costs may contribute as much as 60 percent of the total selling price of the house. This means that site supervision is responsible for a product of exceptionally high value when compared to supervisors in other manufacturing businesses. Obviously, recruitment of top-grade, competent persons for each site supervisory position is most important.

Foundation field studies have shown that rigid controls of onsite labor, materials, methods, and equipment contribute heavily to a successful building operation. According to many successful builders, such controls can be worth up to 5 percent of the selling price of the house.

Preplanning

In the "Production Planning and Control" section of this manual, the importance of scheduling is emphasized. A schedule, either complex or simple, is very important in assigning work crews, ordering materials, and placing major equipment. However, planning should not stop with the schedule.

The day-by-day prethinking of work to be done by individual workmen or crews of workmen can be an important source of productivity improvement. This may be called microscheduling. Such planning is usually left to the leadman or working foreman with some coordination from the site superintendent. Microscheduling is usually less formal than the master schedule and may be communicated orally rather than by a written procedure.

Preplanning is the process of anticipating and correcting material shortages, interference with other crews, equipment delays, or other pitfalls before they cause problems. Preplanning is also the assignment of work to individuals according to their skills. For example, a highly skilled journeyman carpenter carrying materials from the stockpile to the work site is counterproductive. The least costly crew member can do this task and free the highly paid workman for tasks requiring his higher degree of skill. The masonry trade appears to use this concept more completely than other construction trades with laborers, mudmen, and masons doing their specialty tasks separately.

Preplanning often identifies inefficient methods simply because the foreman thinks the entire job through beforehand. For example, while planning the day's work, the foreman visualizes a recurring problem that, in effect, idles four men in a five-man crew for a period of time, such as layout work, if it is not done in advance. His new plan might be to schedule tasks so that the problem will occur at a more opportune time for reassignment of work. Or, better yet, he may be able to reduce or eliminate the problem altogether. If the problem is in house design, he can discuss the extra cost in lost time of the design feature with his supervisors. It may be that the value of the design feature is negligible when the true costs are known. If the problem is one of improper tools or equipment, the foreman can give the superintendent important data for making a cost/benefit study for purchasing more efficient tools or equipment.

Most other industries have a somewhat controlled or constant worksite and work stations within that site. In construction, worksites often tend just to "happen" and grow topsy turvy without much planning or coordination. Because the site often appears to be somewhat disorganized, the individual work crews are often disorganized in their attitude. Preplanning can help organize the job for most efficient use of manpower and equipment.

A clean, well-organized site also leads to better preplanning, which often breaks down when materials are scattered, when equipment cannot be found, or when workmen are forced to climb over construction site clutter. Controlling scrap and waste material makes workers recognize the value of materials.

Crew Composition and Size

Crew composition is a vital part of labor cost control at the building site. Crew size and skill requirements should be evaluated to determine the best combination of workmen to do the work within the respective construction schedule. Ideally, a well-trained one-man crew is most efficient—because he cannot interfere with himself. If a one-man crew is impractical, as in most cases of rough carpentry operations, crew members can be taught to work independently except for those tasks that require cooperative effort.

Previous Foundation field studies reveal that small crews are generally more productive than large crews. Elements of both support labor and nonproductive time, such as excessive material handling, avoidable delays, excessive coffee and lunch breaks, idle time, rework, and so on, are more likely to occur in large crews.

Crew composition also varies from site to site, regardless of crew size. The type of work done by journeymen, apprentices, or laborers varies even on similar jobs. In many cases, workers are overskilled for many of the tasks they perform. For example, if apprentices and laborers are able to complete successfully a task as effectively as a journeyman, then journeymen are considered overskilled for that task and might be used better elsewhere. Whenever a journeyman works on tasks that can be done equally well by an apprentice, labor costs are higher than necessary, and, in dollar terms, productivity decreases.

More than 87 percent of all rough carpentry work requires semiskilled or low-skilled workers. Accordingly, a framing crew composed entirely of skilled journeymen is overskilled, and the employer is paying more than is necessary.

In some areas of California, rough carpentry is specialized to the degree that floors are framed by one small crew, subfloors applied by another crew, walls framed by a third crew, etc. This method appears to be more productive (at the site) than the practice of one large crew framing the entire house. However, tight scheduling plans are required for this type of specialization. In addition, enough volume is needed to permit workmen to move from house to house without delay.

One Ohio builder, who had been using two six-man rough carpentry crews for a number of years, was not satisfied with productivity even though all workmen were relatively skilled and conscientious. He experimented with different crew sizes until he found that three three-man crews produced as much as the two six-man crews. This created a 25 percent reduction in labor costs. It also improved morale because it created more leadman positions.

Industrialization of residential construction often is considered by the uninformed to be offsite component construction or modular housing. Most builders, however, have initiated a substantial amount of onsite industrialization. Material handling equipment and power tools are used widely in some areas of the country. Central cutting yards and small onsite component fabrication areas have been successful for some builders. The assembly-line concept has been applied on larger tracts, especially in areas that tended to use specialists. In these areas, much thought was given to crew size insofar as it affected the industrialization approach. For example, on one site, the builder had determined that the elapsed time for framing a floor with three men was about the same as the elapsed time for applying the subfloor with a two-man crew. Therefore, three floor-framing specialists and two subfloor specialists were used. In this manner, the subfloor crew followed the framing crew from house to house with very little lost time because of interference. Because of repetition, carpenters on these specialist crews became quite productive.

On sites where construction is more closely related to sales, where volume does not warrant use of specialists, where design is complicated, or where houses are built on scattered lots, the tendency is to use more general "all-around" carpenters and fewer production aids, for example, handling equipment and power tools. Under these conditions, crew sizes tend to be larger.

Although smaller crews usually have less nonproductive time than larger crews, exceptions do happen. A study of a seven-man framing crew resulted in about the same percentage of nonproductive time as several two-man framing crews. The study showed, however, that the seven-man crew was actually three two-man crews and one leadman. The carpenters paired off and each resultant two-man crew worked independently of the others. The leadman worked alone until advice or help was needed. Very little interference was observed because work was preplanned.

In conclusion, small crews are usually more productive than large crews. The exceptions, as mentioned above, show no hard and fast rule that large crews are less productive if the workload is preplanned and individual crew members are assigned separate tasks.

The table, *Recommended Crew Size and Composition for Residential Construction Operations*, suggests makeup of crews for different construction operations. The suggested crew sizes have been developed from many onsite studies. For some builders, the recommended crew sizes and composition may be impractical for any number of reasons. However, productivity may be increased and overall labor costs decreased in many situations by using the numbers in this table as guidelines.

Rework

Labor quality is usually more variable than any other part of construction. Some facets of quality, or lack of it, can be quite evident in the finished product. Thus, quality control becomes an essential part in controlling rework, in preventing callbacks, and in ultimately satisfying the customer.

Quality means different things to different people. What quality construction is to one is not necessarily the same to another.

Recommended Crew Size and Composition for Residential Construction Operations

Construction Operation	Type Construction	Crew Size	Composition	Remarks
Carpentry				
1. Floor Framing	One-story crawl space	2	1 Carpenter 1 Apprentice	
	One-story basement	3	1 Carpenter 2 Apprentices	
	Two-story crawl space and basement	3	1 Carpenter 2 Apprentices	
2. Plywood Subfloor	One- and two-story crawl space and basement—all hand nailing	2	2 Apprentices	Assuming relatively simple plywood layout. If complicated, 1 carpenter and 1 apprentice.
	One- and two-story crawl space and basement—power nailing	2	2 Apprentices, position and tack	Assuming relatively simple plywood layout. If complicated, 1 carpenter and 1 apprentice.
		1	1 Apprentice, power nail	
3. Exterior Wall Framing	One- and two-story frame only (no sheathing) up to 24' walls.	2	1 Carpenter 1 Apprentice	24' framed exterior wall will weigh about 200 lbs. May need help in tilt up.
	One- and two-story frame only (no sheathing) up to 48' walls	3	1 Carpenter 2 Apprentices	48' framed exterior wall will weight about 400 lbs. May need help in tilt up.
4. Interior Partitions	One- and two-story	2	1 Carpenter 1 Apprentice	If two-story, two 2-man crews can work independently on different floors.
5. Roof Framing	One- and two-story conventional rafter framing	3	2 Carpenters 1 Apprentice	
	One- and two-story roof trusses, up to 5/12 pitch, 30' length, assistance from lifting equipment	3	1 Carpenter 2 Apprentices	All 3 men top side. (Note: On one study, a two-man crew set trusses with mechanical assistance.)

Recommended Crew Size and Composition for Residential Construction Operations

Construction Operation	Type Construction	Recommendations Crew Size	Composition	Remarks
Carpentry (Cont.) Roof Framing (Cont.)	One- and two-story roof trusses, up to 5/12 pitch, 30' length, no mechanical assistance	4	1 Carpenter 3 Apprentices	2 men top side, 2 men on ground. Fasten one end and ridge when tilting up, then fasten other end. (Note: One study proved 3 men could do work if all trusses were lifted manually before being set.)
	One- and two-story roof trusses, over 5/12 pitch and 30' length, assistance from lifting equipment	4	1 Carpenter 3 Apprentices	Trusses cumbersome and heavy and require extra muscle power to tilt up.
	One- and two-story roof trusses, over 5/12 pitch and 30' length, no mechanical assistance	5 or 6	2 Carpenters 3 or 4 Apprentices	Crew size depends on truss pitch and length. Much nonproductive time can be expected on steeper, longer trusses.
6. Roof Sheathing	One- or two-story, assistance from lifting equipment, hand nailing	2	2 Apprentices	Assumes relatively simple plywood layout. If complicated, 1 carpenter and 1 apprentice.
	One- or two-story, no mechanical assistance, hand nailing	3	3 Apprentices	If complicated layout, 1 carpenter and 2 apprentices.
	One- or two-story with assistance from lifting equipment, power nailing	2	2 Apprentices, position and tack	If complicated layout, 1 carpenter and 1 apprentice to position and tack, 1 apprentice to power nail.
		1	1 Apprentice, power nail	
7. Exterior Wall Sheathing (Not applied while framing)	One-story plywood, insulation board, etc., no scaffolding	2	1 Carpenter 1 Apprentice	
	Two-story plywood, insulation board, etc., scaffolding	3	1 Carpenter 2 Apprentices	2 men on scaffold, 1 below cutting and handling material.

Recommended Crew Size and Composition for Residential Construction Operations

Construction Operation	Type Construction	Crew Size	Composition	Remarks
Footings and Foundation				
1. Dig for footings	Crawl space or slab with trencher	2	1 Operator 1 Laborer	Laborer cleans trench by hand.
	Crawl space, slab or basement, with backhoe	3	1 Operator 2 Laborers	Laborers square, level, and clean trench by hand.
	Crawl space, slab or basement, by hand	3	Laborers	
2. Form for footings and foundation wall	Crawl space or slab, wood or steel forms up to 4' high	2	1 Carpenter or Mason 1 Laborer	
	Crawl space or basement, wood or steel forms, over 4' high	4	2 Carpenters or Masons 2 Laborers	
3. Pour concrete footings	Crawl space, slab, or basement	2	1 Finisher 1 Laborer	
4. Pour concrete foundation wall	Crawl space or slab up to 4' high wall	2	1 Finisher 1 Laborer	
	Crawl space or basement, over 4' high wall	4	2 Finishers 2 Laborers	
5. Concrete block foundation wall	Crawl space, slab, or basement	5	3 Masons 2 Laborers	Laborers arrive early to prepare mortar and distribute block.
Plumbing				
1. Underground DWV and supply	Crawl space, slab, or basement	3	1 Plumber 1 Helper 1 Laborer	Laborer digs trench and backfills.
2. Above ground DWV and supply, rough plumbing	Crawl space, slab, or basement	2	1 Plumber 1 Helper	Working on different systems when possible.
3. Finish plumbing	Crawl space, slab, or basement	1	1 Plumber	

Recommended Crew Size and Composition for Residential Construction Operations

Construction Operation	Type Construction	Recommendations Crew Size	Recommendations Composition	Remarks
Electrical Wiring				
1. Rough wiring	Crawl space, slab, or basement	2	1 Electrician 1 Apprentice	Work on different parts of electrical system.
2. Finish wiring	Crawl space, slab, or basement	1	1 Electrician	
Interior Painting				
1. Spray painting	Gypsumboard or plaster walls, wood trim	1	1 Painter	Two or more 1-man crews can work in different rooms of same house.
2. Brush and roller paint	Gypsumboard or plaster walls, wood trim	2	1 Painter 1 Apprentice	1 man roller painting, 1 man brush painting trim and touch-up.
Drywall				
1. Hanging	Gypsumboard or panelling	2	1 Hanger 1 Helper	
2. Finishing	Gypsumboard	1	1 Finisher	Two or more 1-man crews can work in different rooms of same house.
Heating, ventilation, and air conditioning	All types	2	1 Journeyman 1 Apprentice	

In one Research Foundation study, the finish carpenter found 7 man-minutes of rework when he inspected his own work, the superintendent found 139 more man-minutes of rework, and the builder found even more items of unsatisfactory quality. Acceptable quality means that some minimum standard has been established for comparative purposes. For this reason, the establishment of a standard specification sheet for each house model should be made available to workers, subcontractors, prospective buyers, and the superintendent to help establish a uniform idea of quality. Even so, quality control is a constant problem. The site supervisor can ensure that quality of construction conforms to the standard specifications as published. This type of information, in conjunction with other controls, will help alleviate excessive rework during construction and after the dwelling is completed.

Many times change orders are not implemented because the subcontractor does not receive the order or because it is received after original work has been completed. Resulting delays cause extensive rework for one or more trades to implement the change order. Many of these can be eliminated by having the site supervisor review the change-order log daily. He should perform followup action by verbal communications to those involved, and of course, the change-order system should provide for written communications to the sub or others as appropriate. The site supervisor should perform visual inspections to see whether the work has been accomplished or is in the process of being accomplished.

Rework often is caused by the lack of communications between management, subcontractors, suppliers, workers, and supervisors. Because the site superintendent is at the job a majority of time, he can do much to bridge the communications gap and improve quality by observing what is going on around him. A well-thought-out quality control program can assist the builder in achieving a significant amount of labor cost control and savings at the site.

Equipment Utilization

"Equipment" means all types of equipment from small hand tools to heavy capital equipment. The intent is to stress the importance of adequate equipment, full utilization of available equipment, and to be sure that equipment is available at the jobsite when the schedule requires it.

Many field studies have revealed a considerable amount of lost time on the job because of inadequate equipment, faulty equipment, misscheduling equipment, or workers searching for equipment. If site supervision had been more alert, lost time in most cases would have been averted or minimized.

The two most important reasons for buying equipment in the first place are to reduce total cost and to improve working conditions at the jobsite. If needed equipment has been furnished, the onsite supervisors are then responsible for seeing that it is used properly. They should see that equipment is at the jobsite as scheduled and that it is utilized to its maximum capacity. Further, someone should be assigned responsibility for maintenance, safe storage, and delivery and put-away of the equipment.

These are some pointers to assist site supervision on the daily tasks of overseeing equipment and equipment utilization:
- Make certain that equipment is available as scheduled.
- Instruct tradesmen of the importance of equipment utilization and maintenance.
- Use the proper equipment for each job.
- Use multipurpose equipment when possible.
- Be on the alert for additional equipment requirements.
- Establish a preventive maintenance program that will extend the life of equipment.
- Reduce the interdependency of men and machines wherever possible.

Methods

Methods analysis is the technique that subjects each operation or given piece of work to close study and evaluation in an effort to eliminate unnecessary operations and to approach the quickest and best method of performance for each necessary operation. By using this type of reasoning or analysis for each phase of the operation, many cost control and saving ideas can be implemented. The following subjects are prime candidates for a methods improvement program:
- Crew sizes.
- Simple labor-saving devices.
- Proper use of equipment.
- Improvement of working conditions.
- Worker training.
- Work method simplification.
- Reduction of unnecessary walking.
- Combination of operations.
- Reduction of idle time.
- Elimination of delays.
- Elimination of rework.

Methods improvement involves the following steps:
- Gather details and facts about the job.
- Analyze details to determine the true nature of the job.
- Develop and implement an improved way to do the job.

This is "Finding the Best Way" of doing things. To gain support for such a program, one builder called his men together and told them the company simply had to become more efficient to improve the security of their jobs and that he had designated one of the supervisors (who was competent and well liked) to work full time for a while trying to find ways to work smarter and easier. He asked them to look also for better methods to save both labor and material. He set up a modest award program for their ideas on improvement and made sure that the award-winning employee got plenty of credit for the idea. This latter item, he believed, was more important than the actual amount of the award. The program worked fine and the men devised numerous methods that reduced and controlled cost better.

Although methods of improvement may sound complicated, many of them can be achieved relatively easily. Here is how to get started:

- Assign the superintendent, one of the supervisors, perhaps an engineer or architect or the design staff, or the builder himself should do the analysis. Whoever does it should know construction, be adept at analysis, and have good rapport with the site crews.
- Look first for trouble spots in operations that represent a high percentage of total costs. Also, look for small, easier to solve problems where improvement can be achieved, understood, and appreciated.
- Find the problems that are easier to solve with less work than others. Work on solving these first. For example, take a look at material deliveries for several operations. Can these be better planned to avoid unnecessary material handling later on?
- After selecting the operation to be studied, choose the study date and time.
- For best results, choose a competent crew that is willing to be observed. Inform the crew why and how you are going to do the study (to improve methods, observe time, make their work easier, etc.) Never try to observe a crew without their knowledge. In addition get their suggestions, they will surely come up with some good ideas.

- *Sample Method Improvement Sheet* can be used to analyze each operation. Copies of these sheets can be prepared to fit a clipboard with as many spaces for steps as the sheet will hold.
- Take several of these sheets depending on the length of the operation, write on a clipboard and use a wristwatch for recording times.
- After arriving at the site, fill in the top part of the sheet.
- Write a simple statement of each step in the operation as it is actually being done by the crew. Examples of a step are: carrying materials from pile to house, laying out an interior partition, framing a wall, etc.
- In addition to the other information, describe briefly anything that appears to be an obvious waste of material and labor. For example, during one step, one workman may do nothing except wait for the crew to complete the operation. This and similar items should be noted in the second column. Continue until the operation is completed.
- Within 1 or 2 days after the field observation and while the operation is still fresh in mind, go over each step in detail.
- After the questions have been answered, use the same form in developing a new step-by-step method. Present the new method to the crew in the field. After the changes have been explained to the crew, conduct a trial run to work out any problems that might arise.
- If the new method shows definite improvement, put it into effect immediately. Remember, the methods improvement program requires action to be effective. Again, the program sounds more complex than it really is. If the company is large enough, the ideal person for the job is a full-time industrial engineering consultant, but be sure that he really knows residential construction. A number of builders have found that a week or a few man-weeks on the job and about an equal amount of analysis and followup time of a really good consultant can pay for his services in less—sometimes much less—than a year.

If the above outline seems too complex or the volume of building is small, just go out on the job and watch, with a very critical eye, various operations, especially those involving the big material and labor tickets. Look for idle time, unproductive walking and material handling, rework, site conditions that make working difficult, equipment problems, overuse of materials, materials misuse, and related problems. Then devise a better way and put it into effect.

Sample Method Improvement Sheet		Date	
Phase		**Prepared by**	
Operation		☐ Present ☐ Proposed	
Break Down the Operation		**Question Each Step**	
Steps Write a Simple Statement of Each Step in the Operation	**Indicate** Distances-Time No. of Men, Etc. Difficult Working Conditions— Hazards Excessive Walking	Why is it necessary? (Eliminate?) Where should it properly be done? (Change place? Combine?) When should it be done most effectively? (Change sequence? Combine?) Who is most suitable to do it? (Reassign duties? Combine?) How can it be done best? (Simplify?)	
1			
2			
3			
4			
5			
6			
7			
8			
9			
10			
11			
12			
13			
14			
15			
16			
17			
18			
19			
20			

Actual Onsite Industrial Engineering Studies

Framing—Example 1

A six-man carpentry crew was observed setting prefabricated wall panels, trusses, and gable ends in Indiana. A large forklift was used to set components in place. This study is an excellent example of how improper crew size affects nonproductive time. A large percent of the total time, 18.3 percent, was spent waiting for others to accomplish work and 13.8 percent was spent in personal, idle, and unauthorized talk time. Even though material handling equipment was used, total time for handling was excessive, because the six men, in order to keep busy, assisted the equipment operator unnecessarily when positioning components.

The workmen on this job were apparently good friends who socialized after work and had worked together for many years. A problem may sometimes occur when supervision becomes lax and workmen become close friends. They tend to discuss their personal life more often than would a crew consisting of relative strangers. Because of their longevity with the builder, the crew had become somewhat inefficient. The builder had a commendable policy of retaining people who had worked for him a number of years. But one elderly crew member who was particularly nonproductive could better have been used on less strenuous tasks such as trim work, clean-up, cabinetry, etc.

Also, the nonworking supervisor on this job appeared to have become primarily an expeditor. The proximity of the builder's warehouse to the jobsite appeared to provide workmen with an excuse for leaving necessary tools and equipment or materials behind, and the supervisor spent most of his time traveling back and forth to the warehouse to get forgotten items. He never appeared to have time to supervise the job properly.

Framing—Example 2

A six-man framing crew, four journeymen and two apprentices, was observed framing exterior walls and roof rafters in Maryland. A precut framing package was used. Nonproductive time amounted to almost 28 percent of the total time, most of which was wait-mate time. This study indicated that crew size and balance are primary factors in productivity improvement.

Much scrap and waste was observed. Also, materials were used inefficiently. Material handling was somewhat excessive because the precut package was not properly stacked and much time was spent sorting and searching for materials. The analyst observed that material problems were related primarily to house design and careless housekeeping. For example, where extra studs were placed under breaks in top plate, four or five studs could have been saved by placing the break over a modular stud or header and using modular dimensioning in the first place.

The framing procedure used on this house was not conducive to maximum output. The layout man had nailed bottom plates to the floor previously and had nailed the two top plates together. The framing crew toe-nailed studs and other components to the top plate, tilted up the walls, and toe-nailed studs to the bottom plate.

Most builders consider toe-nailing an archaic method of framing walls. Past industrial engineering studies prove that end-nailing is 7 percent faster than toe-nailing wall framing members.

Framing—Example 3

A three-man crew was observed installing floor joists and plywood subflooring in Indiana. Idle time was excessive for no apparent reason indicating a lack of supervision. With better supervision, idle time could probably have been reduced by 75 percent. A steel floor joist system was used with a precast concrete grade beam. Insulation was installed around the grade beam, joist hangers were placed four feet on center, and joists were set into the hangers. Small adjustable jacks were placed at the center of each joist. The subfloor was 1-1/8-inch tongue-and-groove plywood glued and nailed with special screw shank nails.

When new systems are introduced, an evaluation of proper crew sizes, handling techniques, equipment, etc., must be made. What is considered proper for one system may be completely improper for another. In this study, material handling and nonproductive time could probably have been reduced by preplanning the installation sequence of the steel floor system.

Concrete Flat Work—Example 4

A series of studies were conducted on installation of concrete slabs, driveways, walks, patios, etc., in Arizona, Illinois, and North Carolina. A large difference in productivity was found. Some jobs were planned with small crews who used efficient equipment (trenchers, tractors, tampers, etc.). Others used very little equipment and crew sizes were typically large and scheduling was seldom adhered to by either the finishing crews or the ready-mix concrete truck. One study showed three finishers waited for over an hour for the concrete truck to arrive. The wait time accounted for over 21 percent of the total time required to place and finish the concrete flat work.

Because the workers in one case were paid comparatively low wages, supervisors believed that delays were not very costly. In the end, however, in-place cost of concrete, where wages were high, was less than in-place cost in the low-wage area. Delays can cause a seemingly inexpensive operation to become an expensive one.

Also in one area, workmen seemed to overwork the concrete and appeared to have little comprehension of how good was good enough. One finisher was observed pouring water on a slab so as to get enough moisture to trowel, although the slab was smooth before the water was added. This indicates that occasional reeducation or training of workmen may be necessary to improve productivity. On the most productive jobs, the workmen had a backlog of work and were not eager to stay on the job too long. Consequently, concrete was not overworked.

Occasionally, to finish out the day, workmen will "make work" by overdoing a job such as concrete finishing. For example, if the final trowelling is finished at 3 p.m. and the end of the work day is 4 p.m., a workman may decide to give it one more trowel rather than start on a new slab. Often there is enough preparation work to be done on the other slabs so that workers could easily find work to do elsewhere. Supervisors should watch for this tendency to make work and give necessary instructions to reduce it. Midafternoons and on Fridays are good times to look especially for make-work time.

The study also showed that concrete placement is more efficient when ready-mix trucks travel counter clockwise around the slab because the truck drivers can observe the operation.

Brick and Block—Example 5

Studies were conducted on concrete block foundation walls and brick veneer exterior walls. Nonproductive time varied substantially among the studies. For example, a six-man masonry crew in one state had 21.5 percent nonproductive time while a four-man masonry crew in another had 1.8 percent nonproductive time. The variances in material handling and nonproductive time appear to be related primarily to the ratio of masons to laborers and the location of initial placement of the brick and block. Material handling is usually done by the laborers. Therefore, the proper ratio of masons to laborers is important for maximum productivity.

To reduce material handling time, bricks and blocks should be placed as near to the work stations as possible. Concrete blocks should be delivered to the center of the foundation when possible. In these case studies, the first part of the day was usually spent preparing mortar and distributing bricks and blocks near the point of use. This was typically done by laborers while masons either stood idle or spent time nonproductively. If laborers arrived at the site at least half an hour before the masons, preparation work would have been accomplished and nonproductive time would have decreased.

Often, a good crew balance consists of three masons for every two laborers. However, this will not hold true for every situation. Supervisors should be aware of the necessity of creating proper crew balances for the most productive effort.

Plumbing—Example 6

An analysis of four plumbing onsite industrial engineering studies of residential plumbing installations provided valuable cost control and reduction information. Crew size ranged from three to five men. Two of the studies were conducted in Indiana where nonproductive time averaged approximately 40 percent. On one of the Indiana studies, a three-man crew installed all drain waste and vent plumbing in a single-family dwelling. On this study, personal and idle time amounted to over 22 percent of the total time. Rework accounted for 10 percent. Apparently, the lack of supervision contributed to these high totals. Talk business time accounted for over 13 percent which indicated that the crew was not familiar with the work and should have had more supervision.

Unless the plumbers are furnished with schematic drawings of the plumbing system, they design the system as they progress, with much talk business, waiting, and idle time.

Because of the many parts and pieces required for a typical plumbing system, plumbers should bring in all necessary materials at one time. Often material handling amounts to a high percentage of the total plumbing job. In one study, almost 28 percent of the time was in material handling. If the plumber had anticipated his needs, he could have reduced travel time to and from the truck considerably and reduced material handling time by at least 75 percent.

A possible solution would be the development of a plumbing checklist, containing a listing of all possible fittings and other materials to be used in a dwelling. Average quantities of each piece required could be developed in a short period of time. In this manner, a plumbing package could be put together offsite for each dwelling unit. A nominal amount of spare pieces could be included in the package for contingencies.

Electrical—Example 7

Two electrical onsite industrial engineering studies, one in the West and another in the East, were analyzed. The western study, in a one-story home, where the installation procedure was unique, is described in some detail. Four journeyman electricians did the wiring, each doing a portion that was considered his specialty. One man drilled all holes, pulled wire and installed boxes. He was followed by another electrician who tied in wires at the panel and at the outlet boxes. A third electrician, considered the plug-and-switch man, did nothing but tie wires to receptacles and switches. A fourth man followed and installed fixtures and cover plates. Each man, therefore, became a specialist in a portion of the electrical work, and in doing so, each became proficient at his specialty. The entire house, including finish trim, was wired in approximately eight manhours. When discussing this method with the electrical subcontractor, specialties of the workmen were even further defined. The subcontractor explained that different men were more efficient not only at different tasks but also in different types of construction. In other words, one plug-and-switch man might work on one-story houses only. Another plug-and-switch man would be more efficient on multistory houses. The electricians were paid relatively high wages for, in some cases, relatively low skill tasks. However, because of the specialist approach, total labor costs for wiring was lower than found elsewhere. This study indicated that this type of specialization can work onsite with proper supervision, planning and proper application of manpower and development of skills.

In the eastern study, electricians worked as generalists within their trade, meaning that they worked on all tasks required for the job.

The lack of repetition meant that the same task might not be repeated for several days on the eastern job. For example, over 35 manhours were required to wire this house. The same series of tasks would occur approximately once a week. In the other, the same series of tasks occurred several times per day for each specialist.

The one-story western house contained 1,650 square feet of floor area and 61 outlets. The two-story eastern house contained approximately 2,000 square feet of floor area and 80 outlets. Unit time comparisons indicated that improvement in productive direct time can be expected by using specialists.

The western study also indicated that tools and equipment are very important in improving productivity. Electricians there had a proper set of tools that fit their personal circumstances. For example, the fixture and plate man had a battery powered screw driver. Because most of his work was screw fastening material together, he found that he could improve productivity with this tool. The percentage of time spent in productive direct effort on the two-story house was not substantially different from that observed on the other, although total time to wire the two-story house was about four times as great. Some of this difference was because of house design. Most of the variance was because specialists versus generalists within the trade made it possible to significantly improve productivity, that is, output per manhour of work.

Other Labor Cost Control and Savings Ideas

All the industrial engineering studies of labor cost control point to one fact. Typically, there is no one place to make large labor cost reductions. On the other hand, the onsite studies show over and over that there are usually opportunities to reduce labor costs a little bit in a lot of places. These small savings soon add up to substantial cost reductions. Applying the labor cost control principles discussed could mean a reduction of 15 percent in labor time, and in some cases, 30 percent is not beyond reach.

To make a lot of little labor savings possible, consider the following checklist.

Use Labor Time Accounting System

Many builders use some kind of time card system; consider using such a system to provide more detailed information about labor times by work categories. Such labor time information has many uses related to increasing control and productivity.

Set Up Work Standards

Builders and industry have shown that unit work standards can be used to increase productivity with or without incentive wage payments. Good standards will not eliminate avoidable delays, idle time, personal time, and excessive break time, but they will cut them to a minimum.

Use Good Work Planning

As previously discussed, good work planning can save a substantial amount of labor. Basically, this means planning in advance who is going to do what, when, and how.

Repeat Work Operations

Assign workmen to as many repeat operations as possible. Repetition almost always improves productivity.

Train Leaders

Good supervision is the key to labor cost control. Crew leaders and supervisors should be well trained in labor and material cost control. Remember, the fastest workman may not necessarily be the best supervisor.

Provide Detailed Drawings

Make detailed drawings to eliminate jobsite confusion and resultant lost time. Show unusual details in clear, large scale so that workers do not have to take time to figure out how "specials" are to be built.

Require Proper Accuracy

Instruct supervisors to concentrate on getting a proper degree of accuracy, where critical, and to avoid high accuracy when it is not necessary. For example, the top of a concrete footing can be leveled with a shovel. It does not need a smooth steel troweled finish. In fact, the bond is better if the surface is somewhat rough and irregular.

Use Proper Tools

Constant checking on the use of the right tool for the right job is very important. For example, in spray painting, an airless sprayer substantially reduces masking labor time.

Because material costs are usually greater than labor costs, labor is very important in the way it uses materials. Supervisors need to understand that wasting materials to save labor is seldom cost effective. Overall, in-place labor and material cost control and reduction cannot be truly separated. In some cases, increasing labor slightly may save materials. For example, if labor is hurried in preparing gravel fill for a slab, more concrete may be required. Extra time spent bringing gravel to the proper level in this case could save several cubic yards of costly concrete.

The process of construction cost control requires a continuing analysis of all facets of cost and their interrelationships.